GLOBAL WATER MONITORING REPORT

CONTINENTAL DRYING

A Threat to Our Common Future

Fan Zhang, Christian Borja-Vega,
Hrishikesh Arvind Chandanpurkar,
James Famiglietti, Rick Hogeboom,
Regassa Namara, Zarif Rasul,
Pavel Luengas-Sierra,
and Deyu Rao

WORLD BANK GROUP

1 2 3 4 28 27 26 25

ISBN (paper): 978-1-4648-2269-8
ISBN (electronic): 978-1-4648-2270-4
DOI: 10.1596/978-1-4648-2269-8

Cover photo: EyeEm Mobile GmbH / iStock

Cover design: Melina Rose Yingling / World Bank Creative Services, Global Corporate Solutions

Library of Congress Control Number: 9781464822698

CONTINENTAL DRYING

Reproducible Research Repository

https://reproducibility.worldbank.org

A reproducibility package is available for
this book in the Reproducible Research Repository at
https://reproducibility.worldbank.org/index.php/catalog/386.

Scan to see all titles in this series.

Global Water Monitoring Report

The Global Water Monitoring Report series provides regular updates on the state of global water resources and highlights priority areas for investment and policy interventions. Each edition offers an in-depth analysis of a critical water and development issue. This series aims to support the World Bank Water Global Department in its operations, policy dialogue, and engagement with partners.

Titles in This Series

2025

Continental Drying: A Threat to Our Common Future, Fan Zhang, Christian Borja-Vega, Hrishikesh Arvind Chandanpurkar, James Famiglietti, Rick Hogeboom, Regassa Namara, Zarif Rasul, Pavel Luengas-Sierra, and Deyu Rao

All books in the Global Water Monitoring Report series are available free at https://hdl.handle.net/10986/43724.

Contents

BOXES

FIGURES

MAPS

TABLES

TECHNICAL APPENDIXES

Technical appendixes A through E are available online at https://hdl.handle.net/10986/43683.

Foreword

Water is life. Yet across continents, this vital resource is vanishing at an alarming rate. Each year, the world loses 324 billion cubic meters of freshwater—the equivalent to the combined annual flow of Western Europe's largest rivers (Danube, Rhine, Elbe, and Meuse), or enough to meet the annual needs of 280 million people. This is no longer a future risk; it is a silent crisis already reshaping economies, ecosystems, and lives.

This first edition of the Global Water Monitoring Report, *Continental Drying: A Threat to Our Common Future*, presents a sobering view. Drawing on two decades of satellite data and advanced modeling, it reveals a clear and troubling trend: global freshwater reserve has declined by 3 percent of the freshwater supply on average, annually, and by as much as 10 percent in already-dry regions. Across the planet, dry areas are getting drier, and wet areas continue to get wetter—but the pace and level of drying are both more widespread and more rapid. This shift in the global water cycle carries severe consequences: jobs lost, incomes diminished, ecosystems degraded, and new vulnerability hot spots emerging, even in places once considered water secure.

The impacts are deeply interconnected, with regional and global ripple effects on jobs, incomes, value chains, and trade. Moreover, freshwater depletion is heightening the risk of wildfire, especially in biodiversity hot spots.

For the first time, we now have a much more precise picture of where and why freshwater is disappearing. A new methodology improves satellite resolution from about 330 kilometers to 25 kilometers, enabling detailed analysis at basin, country, and even county levels. Combined with economic and land-use data, this technology reveals the underlying drivers: global warming, intensifying droughts, and unsustainable land and water practices.

This report goes beyond diagnosis—it is a call to action with a roadmap for possible solutions. It proposes a three-pronged approach: managing demand, augmenting water supply, and improving water allocation to tackle the continental drying crisis. Five cross-cutting levers are key to success: strengthening institutions, reforming tariffs and repurposing subsidies, adopting water accounting, leveraging data and technological innovations, and valuing water in trade.

Together, these measures can accelerate implementation and attract private investment to finance the transition. Grounded in international best practice, these policy options can help reverse the trend of increased continental drying. With the right knowledge, partnerships, and political will, we can turn the tide.

Axel van Trotsenberg
Senior Managing Director for Development and Policy
World Bank

Acknowledgments

The first in the Global Water Monitoring Report series, *Continental Drying: A Threat to Our Common Future,* was prepared by a team led by Fan Zhang and comprising Christian Borja-Vega, Hrishikesh Arvind Chandanpurkar, James Famiglietti, Rick Hogeboom, Regassa Namara, Zarif Rasul, Pavel Luengas-Sierra, and Deyu Rao. The book greatly benefited from the strategic guidance and general direction of Axel van Trotsenburg (Senior Managing Director), Juergen Voegele (Vice President, Planet Vice Presidency), Renaud Seligmann (Director, Strategy and Operations, Planet Vice Presidency), and Saroj Kumar Jha (Global Director, Water Department).

Additional World Bank contributors include Patrick Kabanda, Poolad Karimi, Amjad Muhammad Khan, Homero Alejandro Paltan Lopez, Ana De Menezes, and Amal Talbi. Additional external collaborators include Markus Berger (University of Twente), H. Ron Chan (University of Manchester), Kaoru Kakinuma (Korea Advanced Institute of Science and Technology), Shanjun Li (Stanford University), Oleksandr Mialyk (University of Twente), Ivan Rudik (Cornell University), Han Su (University of Twente), and Yoshihide Wada (King Abdullah University of Science and Technology).

The team is grateful to colleagues from the World Bank Group and external contributors for their inputs and suggestions at various stages of the book's development, including Soumya Balasubramanya, Eileen Burke, Anna Cestari, Doerte Doemeland, Alexander Fergusson, Francis Ghesquiere, Sumila Gulyani, Michael Haney, Nagaraja Rao Harshadeep, Stephen Hodgson, Elizabeth Huybens, Fatouma Ibrahima, Christina Leb, Abel Mejia, David Michaud, Sebasitian Molineus, Victor Mosoti, Soma Ghosh Moulik, Yogita Mumssen, Sarah Nedolast, Daniel Nolasco, Francois Onimus, Martin Raiser, William Rex, Dena Ringold, Gustavo Saltiel, Yitbarek Tessema, Michael John Webster, and Jason Zhang; and Peter Debaere (University of Virginia), Dustin Garrick (University of Waterloo), Sujay Kumar (National Aeronautics and Space Administration Goddard Space Flight Center), John Reager (California Institute of Technology Jet Propulsion Laboratory), Matthew Rodell (National Aeronautics and Space Administration Goddard Space Flight Center), and Dale Whittington (University of North Carolina).

The team also extends its gratitude to peer reviewers for their invaluable feedback and advice, including Luis Andres (Lead Economist, World Bank), Benoit Bosquet (Regional Practice Director, World Bank), Ranveer Chandra (Managing Director, Microsoft), Richard Damania (Chief Economist, Planet Vice Presidency), Guillermo Donoso (Professor, Universidad Católica de Chile), David Mutombo (CEO of Regideso, DRC), Hurun Onder (Associate Professor, Texas A&M University), Dan Vardi (Global Lead for Water, IFC), Winston Yu (Practice Manager, World Bank), and Wenhe Zhang (Senior Underwriter, MIGA).

The team thanks Melissa Edeburn, Karen Schneider, Kathie Porta Baker, and Honora Mara for editing and proofreading, as well as Melina Rose Yingling, World Bank Global Corporate Solutions, for cover design. The team appreciates the valuable guidance and support from the World Bank communications team on communications and outreach, including Nigina Alieva, Erin Barrett, Ada Calderon, Sarah Farhat, and Kimberly Versak. The team is grateful for the World Bank's publishing team, including Carmen Fernandez, Diane Kim, Shuanni Lee, Jewel McFadden, and Caroline Polk, for managing, editing, designing, typesetting, printing, and disseminating the print and digital versions of the report. The team is also grateful for the administrative support from Yasmin Angeles, Alberto Arbesu Cardona, Belinda Asaam, Georgine A. Badou, and Marie-Adele Tchakounte Sitchet.

The development of this report was made possible through financial support from the Global Water Security and Sanitation Partnership (GWSP). GWSP is a multidonor trust fund administered by the World Bank's Global Department for Water and supported by Australia's Department of Foreign Affairs and Trade; Austria's Federal Ministry of Finance; Denmark's Ministry of Foreign Affairs; the Netherlands' Ministry of Foreign Affairs; the Gates Foundation; the Spanish Ministry of Economy, Trade and Business; the Swedish International Development Cooperation Agency; Switzerland's State Secretariat for Economic Affairs; the Swiss Agency for Development and Cooperation; and the United Kingdom Foreign, Commonwealth and Development Office.

Main Messages

This report addresses the escalating global water crisis characterized by *continental drying*—the long-term decline in freshwater availability across large landmasses—highlighting trends, causes, and impacts on jobs, the economy, and the planet. It presents a comprehensive strategy and policy action road map focused on demand management, supply augmentation, and improved water allocation to address the water crisis.

Trend and causes of continental drying

Global freshwater reserves have significantly declined over the past two decades, leading to the emergence of mega-drying regions. This loss of freshwater is estimated at 324 billion cubic meters per year—enough to meet the annual water needs of 280 million people. Regionally, the median basin-level annual decline in freshwater reserves is about 3 percent of the annual renewable freshwater supply across all basins but reaches 10 percent in arid regions experiencing drying. Global warming, worsening droughts, and unsustainable land and water use are major contributors to this trend.

Impacts on jobs, the economy, and the planet

Continental drying severely affects agricultural productivity, leading to job losses, income declines, and environmental damages.

- *Jobs.* In Sub-Saharan Africa, droughts have rendered 600,000–900,000 people jobless annually. The negative effect of water scarcity on jobs is most pronounced in rural farming communities and among women, older individuals, landless farmers, and low-skilled workers.
- *Income.* Local water shortages can have global economic repercussions because of interconnected trade networks. For example, a 100 mm drop in annual rainfall in India could reduce global real income by US$68 billion.[1]
- *Wildfires and biodiversity.* Continental drying increases the frequency and severity of wildfires, posing a threat to biodiversity. A 1-standard-deviation (SD) increase in the rate of freshwater depletion raises the likelihood of wildfires by 27 percent; in biodiversity hot spots, it raises the likelihood by 50 percent.

Mapping water vulnerability and savings potential

Global water use increased by 25 percent from 2000 to 2019, with about a third of this increase occurring in regions already drying out. Regions facing the dual crisis of increasing water demand and decreasing supply are identified, with a significant share of water consumption in these drying regions linked to low water use efficiency in the cultivation of water-intensive crops.

Enhancing water use efficiency in agriculture could save substantial amounts of water when paired with effective monitoring and regulations to protect the savings. Globally, aligning the production of key crops with median levels of global water use efficiency could reduce consumption of freshwater from rivers, lakes, and aquifers by 137 billion cubic meters—equivalent to the annual water needs of 118 million people. Adjusting cropland distribution to better match water availability within national borders and reallocating water use from less efficient to more efficient producers and from water-scarce to water-abundant areas across countries through virtual water trade can further increase water savings in drying regions.

Policy recommendations

The report recommends a three-pronged approach: managing demand, augmenting water supply, and improving water allocation to tackle the continental drying crisis. Five cross-cutting levers—strengthening institutions, reforming tariffs and repurposing subsidies, adopting water accounting, leveraging data and technological innovations, and valuing water in trade—are essential for effective implementation and to attract private investment to finance the approach. Beyond water, addressing trade barriers, investing in education and skills development, and improving access to markets and financial services are critical for strengthening job and livelihood resilience.

Note

1. It should be noted that this finding is based on a modeled scenario and reflects the potential global cost of water scarcity in India, rather than a direct measurement of current annual losses.

Overview

Introduction

The year 2024 marked another year of record-breaking water-related extremes across the globe. Widespread droughts and floods not only cause devastating local impacts but also signal broader shifts in the global water cycle (GCEW 2024). A changing water cycle directly affects the availability of freshwater. Although shifts in water extremes are increasingly visible, the long-term trend for total freshwater availability on land remains poorly understood. Conventional wisdom assumes that the amount of freshwater available for human and ecosystem use is fixed and that periodic deficits get eventually compensated with surplus precipitation. However, this report's findings challenge that assumption. They reveal a crisis not only of water extremes but also of a shifting baseline in global water availability.

Focus of this report

Grounded in new evidence from satellite data, this report presents the first global assessment of freshwater reserves over the past two decades. The findings expose an alarming trend of continental drying: a persistent long-term decline in freshwater availability across vast landmasses. Not only are droughts and deluges becoming more unpredictable, but the total amount of freshwater available for use has also significantly declined. Combining satellite data with microeconomic data, this report further explores four questions.

What has caused continental drying?

Previous research and this report's analysis point to three major causes: global warming, worsening droughts, and human activities that affect water and land use. Among these, unsustainable water and land use are pivotal

---

A reproducibility package is available for this book in the Reproducible Research Repository at https://reproducibility.worldbank.org/index.php/catalog/386.

drivers of freshwater loss in nonglaciated regions. Although the crisis of a changing climate is at its core a water crisis, ineffective water management has further exacerbated it.

What are the impacts of continental drying?

Continental drying raises profound risks: It can reduce agricultural productivity, harm human health, and, in extreme cases, render land uninhabitable and drive large-scale emigration. The report reveals new empirical evidence showing how freshwater depletion jeopardizes jobs and incomes, intensifies the frequency and severity of wildfires, and poses a serious threat to biodiversity. In the long term, the combined effects of drying and warming could push societies toward a tipping point at which damage accelerates rapidly and adaptation becomes increasingly difficult.

Where are the water vulnerability hot spots and savings opportunities?

Regions experiencing the fastest decline in freshwater reserves and a substantial increase in water consumption are emerging as vulnerability hot spots. The pressure is compounded by inefficient water use in farming: About one-quarter of inefficient rain-fed and one-third of inefficient irrigated crop water consumption occur in areas already experiencing shrinking water availability. Although these areas are highly vulnerable to long-term freshwater depletion, they also have significant potential to enhance resilience through sustainable water management. Globally, aligning the production of key crops with median levels of global water use efficiency could lead to an 18 percent reduction in irrigation water use in drying areas.

How can the continental drying water crisis be tackled?

This report offers a comprehensive policy blueprint to tackle the water crisis through actions both within and beyond the water sector. Within the water sector, countries must advance comprehensive water reforms that address water demand, supply, and allocation. These reforms include strengthening institutions to enable robust monitoring, enforcement, and cross-sector coordination; smarter pricing systems that reflect the scarcity and value of water; water accounting, which is essential to understand water endowments and risks; innovative data and technology systems that track water availability and use in real time; and trade policies aligned with water sustainability. Beyond the water sector, addressing trade barriers, investing in education and skills development, and improving access to markets and financial services are essential to build more adaptive and resilient economies that can be better prepared to withstand the impacts of water scarcity.

What's new in this report

This report uses perhaps the most comprehensive hydrologic variable—terrestrial water storage (TWS)—to assess changes in freshwater availability over the past two decades. TWS is the total amount of water stored on land. It includes water found in glaciers, rivers, lakes, reservoirs, and aquifers and water held in soil moisture and vegetation. Monitoring changes in TWS—especially water stored beneath the ground, which represents 97 percent of Earth's unfrozen freshwater—is notoriously difficult. That situation changed in 2002 with the launch of the Gravity Recovery and Climate Experiment (GRACE) missions by the U.S. National Aeronautics and Space Administration and the German Aerospace Center. By tracking changes in Earth's gravity caused by shifting water storage, GRACE has made it possible to measure changes in freshwater reserves on a global scale (refer to box O.1).

BOX O.1

Terrestrial water storage: What it is and how it is tracked from space

Think of land as a giant water bank (refer to figure BO.1.1). It holds water in different accounts: short-term stores, such as soil moisture, seasonal snow cover, and small reservoirs (checking account); medium-term stores, such as groundwater in shallow aquifers and large lakes (savings account); and long-term stores, such as groundwater in deep aquifers, glaciers, and ice caps (trust fund). Deposits come from rainfall and snow, and withdrawals occur through evaporation, runoff, and water usage. If withdrawals consistently exceed deposits, the system runs at a deficit, depleting long-term reserves such as groundwater.

Just like a bank account statement, the level of TWS indicates whether regions are saving water, breaking even, or heading into water bankruptcy. Traditionally, however, there was no way to regularly monitor this important variable at all locations on land. That situation changed in 2002 with the launch of the Gravity Recovery and Climate Experiment (GRACE) satellite mission of the U.S. National Aeronautics and Space Administration and the German Aerospace Center.

(continued)

BOX O.1

Terrestrial water storage: What it is and how it is tracked from space *(continued)*

FIGURE BO.1.1 Terrestrial water storage: Earth's water bank

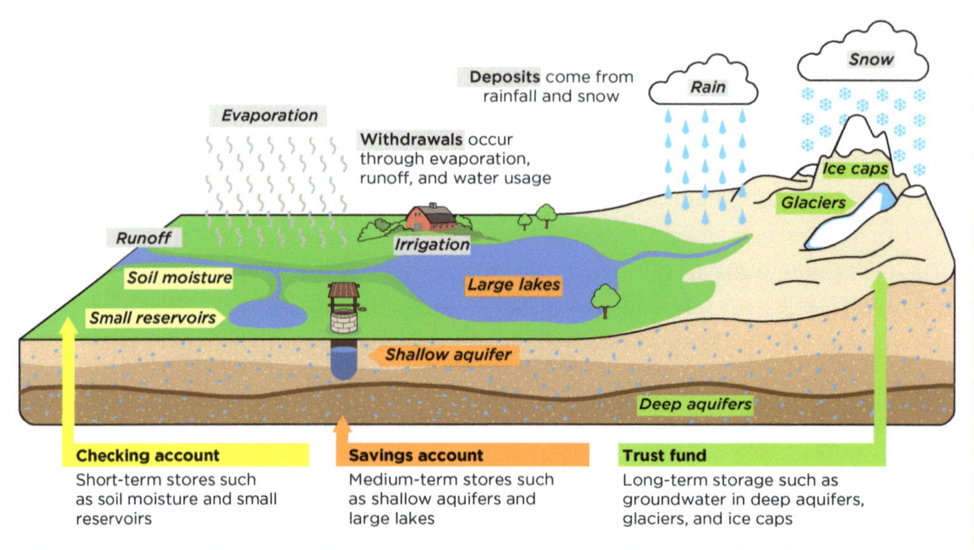

Source: World Bank.

The GRACE (2002–17) mission and its successor, the GRACE Follow-On (2018–present) mission, consist of two satellites orbiting Earth. The distance between them is measured with high precision, down to micrometers (a *micrometer* is 1/50th the width of a human hair). As the pair of satellites circle Earth, their positions are changed slightly by the amount of water mass in the region below. For example, when the front satellite approaches a large water reservoir, it accelerates because it is pulled slightly by the reservoir's gravity, changing the distance between the two satellites. When it has passed over the reservoir, it decelerates and the distance between the two satellites changes again. The next month, if the reservoir has lost water, the overall pull by the reservoir will be smaller; thus, the satellites' acceleration and deceleration will be smaller. The difference between the two months' measurements is then converted into water mass units, after accounting for other influences. By monitoring the small but continuous changes in Earth's gravity field caused by water movement, GRACE functions like a scale in the sky, enabling researchers to observe fluctuations in TWS.

Many studies have analyzed TWS changes using GRACE data (for example, Reager et al. 2016; Rodell et al. 2018, 2024; Rodell and Li 2023; Scanlon et al. 2022). Although existing studies characterize regional trends relatively well, uncertainties remain regarding whether these trends are robust; what the underlying drivers are; and whether the global continents, as a whole, are gaining or losing freshwater from or to oceans (Kim et al. 2019).

Additionally, most studies are conducted at close to GRACE's native resolution, approximately 330 km at the equator, roughly the size of a small country. Although this resolution captures broad patterns, it is too coarse to detect finer-scale changes, such as those occurring in smaller watersheds or local aquifers.

Finally, although continental drying affects vast areas of global land, its impacts are shaped by local water consumption patterns. The most recent global assessment of water consumption, also known as a *water footprint*, was conducted by Hoekstra and Mekonnen (2012), providing 10-year annual averages for 1996–2005. Given significant shifts in water use since then, an updated water footprint estimate is needed to identify vulnerability hot spots facing the dual pressures of declining supply and rising demand.

This report introduces three key innovations to address these knowledge gaps. First, it provides a comprehensive assessment of the long-term global trends for TWS from April 2002 to April 2024. It examines how reliable the trends are by accounting for uncertainties in the satellite data and testing how consistent these patterns are over time, examining, for example, whether areas showing drying today were also drying in earlier years. Drying trends that appear in 95 percent of possible time periods for a region are considered robust. The report also compares long-term changes with year-to-year natural fluctuations (such as those from El Niño) to highlight areas where the drying signal is stronger than natural variability. These checks help identify the most significant and persistent drying regions—those experiencing persistent water deficits year after year.

Combining GRACE data with a global hydrological model, this report also breaks down the change in water storage into its key components—groundwater, soil moisture, surface water, and snow—at the global scale. This decomposition provides insights into the sources of water loss.

Second, this report uses a novel method to enhance GRACE's resolution to approximately 25 km, allowing for more precise characterization of TWS changes at a finer scale, ranging from national and state levels and even to individual counties. These high-resolution TWS data are used to assess the significance of changes in water storage at the local basin level, that is, to measure how much a basin is running a water "overdraft" relative to its annual rainfall deposits.

The downscaled TWS data are further integrated with microeconomic data to identify the drivers and impacts of TWS depletion. Connecting TWS trends with land use and water management indicators provides new empirical evidence on the human drivers of continental drying. In addition, despite growing evidence of the social and economic impacts of water scarcity, the empirical evidence on the environmental and long-term economic impacts of drying, and on whether adaptation can mitigate these effects, is still limited. Using a rich set of high-resolution data, this report helps fill that gap through novel econometric methods. Moreover, the data allow for analysis of regional variations in the impacts of drying, shaped by local infrastructure and institutional conditions, and provides insights into potential adaptation strategies to mitigate these impacts.

Third, this report estimates the water footprint for 175 crops in 169 countries in 2000 to 2019 at high resolution (approximately 10 km), using a state-of-the-art crop growth model and the most up-to-date, globally comprehensive agriculture data. These data are then overlaid with TWS data to identify vulnerability hot spots where water availability is declining, demand is increasing, water use efficiency is low, and a substantial amount of *virtual water*—water embedded in agriculture and water-intensive industrial goods—is used for export rather than domestic consumption. Understanding water consumption patterns is critical for identifying sources of vulnerability and solutions for water savings. This report estimates crop-specific water use efficiency to assess the potential for water savings in drying regions.

The following sections of this overview summarize the report's main findings and recommendations.

Trends and causes of global freshwater decline

A key finding of the report is the persistent net loss of global freshwater availability at an annual rate of 324 billion cubic meters—enough to meet the annual water needs of 280 million people—over the past two decades (refer to figure O.1). Regionally, the median basin-level freshwater loss equals about 3 percent of the annual renewable freshwater supply.[1] In basins that are arid and experiencing drying, the median freshwater loss reaches 10 percent, exacerbating water scarcity in areas that can least afford it.

Global freshwater availability has persistently declined over the past two decades.

FIGURE O.1 Global freshwater loss to oceans, 2002–24

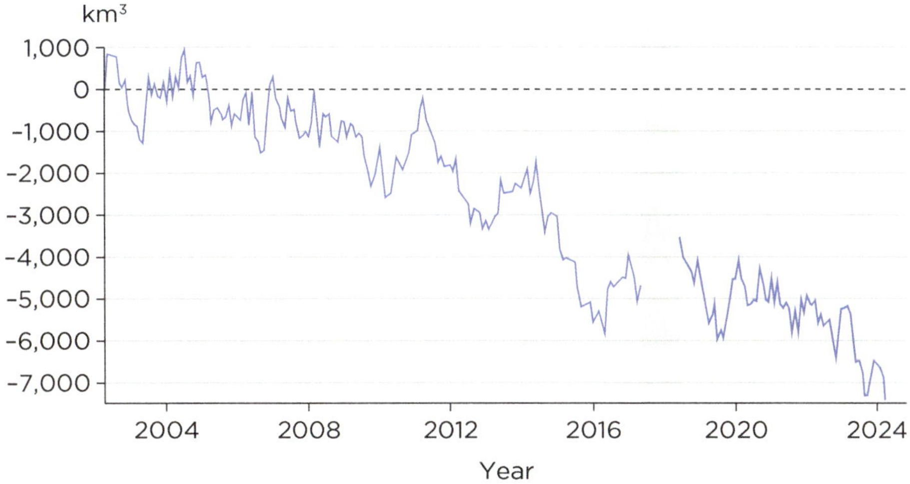

Source: Adapted from Chandanpurkar et al. 2025.
Note: This figure depicts the long-term loss of the global land water mass, measured by the net change in globally integrated, deseasoned TWS data from April 2002 to April 2024. TWS data are from the GRACE satellite mission, which operated from 2002 to 2017, and its successor mission, the GRACE-FO, launched in 2018. The gray band indicates the gap between the GRACE and GRACE-FO missions. GRACE = Gravity Recovery and Climate Experiment; GRACE-FO = Gravity Recovery and Climate Experiment Follow-On; km^3 = cubic kilometers; TWS = terrestrial water storage.

Continental drying as seen from space

Beyond the global average, the trend varies across regions. Although most of the world's dry areas continue to get drier and the wet areas continue to get wetter, the extent of drying is both more widespread and more rapid than that of wetting. The rapid expansion of dry areas has led to the emergence of continental-scale mega-drying regions formed by the convergence of previously identified drying hot spots, as shown in map O.1. These regions include Alaska and western and northern Canada, Central America and southwestern North America, the northern Russian Federation, and the vast landmass including Central Asia, northern China, Europe, the Middle East and North Africa, South Asia (except Peninsular India), and Southeast Asia.

Rapid expansion of dry areas has led to the emergence of continental-scale mega-drying regions.

MAP O.1 Emergence of continental-scale mega-drying regions

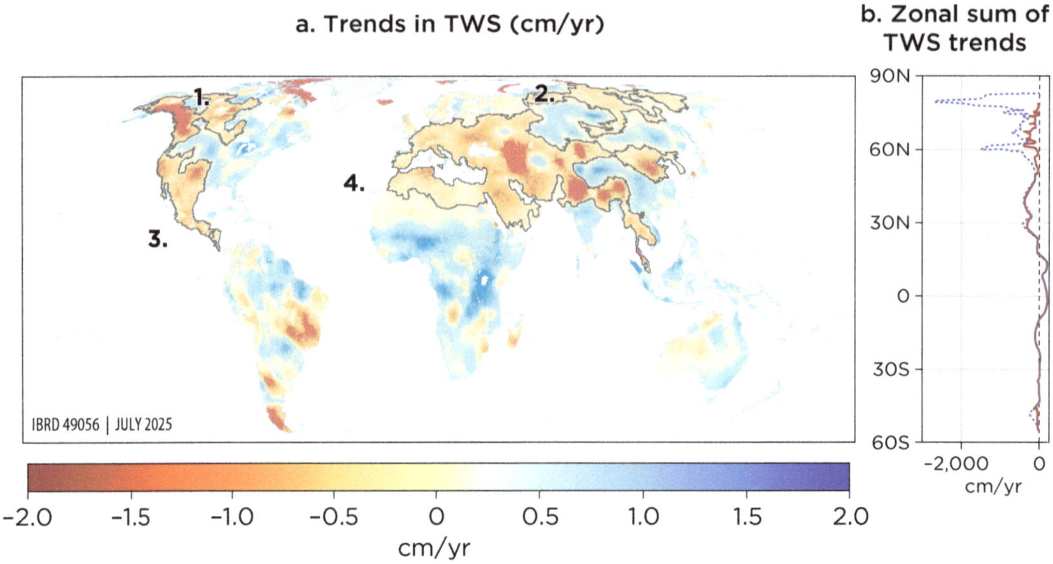

Source: Chandanpurkar et al. 2025.
Note: In panel a, red indicates regions that have experienced a persistent loss of freshwater reserves over the past two decades, and blue indicates areas that have consistently gained freshwater reserves during the same period. Intensity of color corresponds to the magnitude of water loss or gain. The numbers on the map indicate the mega-drying regions: 1 = Alaska and western and northern Canada; 2 = northern Russian Federation; 3 = Central America and southwestern North America; 4 = Central Asia, northern China, Europe, the Middle East and North Africa, South Asia (except Peninsular India), and Southeast Asia. The zonal plot in panel b shows the sum of TWS across parallel lines, for all regions (blue dotted line) and for nonglaciated regions (red). It shows that, with the exception of the tropics between 10S and 20N, all latitudes exhibit a net decline in freshwater reserves, even when excluding glaciers and ice caps. cm/yr = centimeters per year; TWS = terrestrial water storage.

Drivers of continental drying

This report estimates that annual freshwater loss in nonglaciated drying continental regions is increasing each year and has now exceeded the loss from melting glaciers and ice caps (excluding Antarctica and Greenland). In nonglaciated drying regions, the single largest contributor to water storage loss comes from the depletion of groundwater (68 percent), followed by surface water (18 percent), soil moisture (9 percent), and snow water (5 percent).

Although snow and glacier melting are primarily influenced by global warming (Immerzeel et al. 2020), many freshwater losses in nonglaciated regions could be reduced through better local water and land management. Indeed, this report shows that land use change is a pivotal driver of freshwater depletion. Regions with healthy forests and wetlands tend to hold larger freshwater reserves, whereas those dominated by urbanization and intensive irrigation face much faster rates of depletion (refer to figure O.2).

Additionally, addressing agriculture water underpricing that leads to overpumping of groundwater can significantly reduce water loss in irrigation-intensive countries (refer to figure O.3, panel a). Integrated water resources management (IWRM) that balances the competing demands of agriculture, urban development, and conservation also plays a vital role in protecting water resources. In countries with weak IWRM (scoring below the 30th percentile), freshwater reserves are depleted two to three times faster than in nations with more effective management, all else being equal (refer to figure O.3, panel b).

Land use decisions are a key driver of continental drying in nonglaciated areas.

FIGURE O.2 Impact of 1 percent change in land use type in 2002 on TWS trends, 2003–24

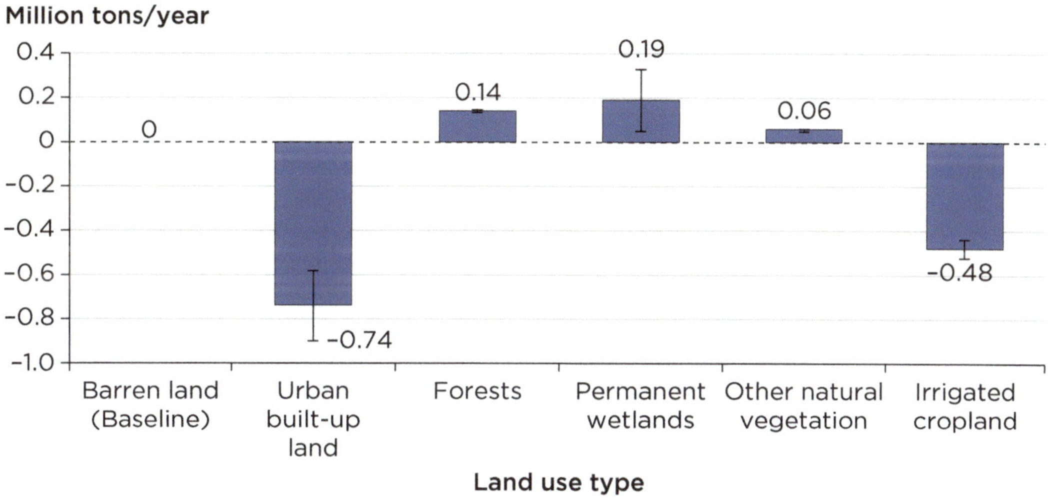

Source: World Bank.
Note: This figure illustrates the estimated impact of a 1 percent change in the respective land use type in 2002 (equivalent to approximately 25 km^2 on average) on the grid cell's TWS trend from 2003 to 2024, with barren land serving as the baseline. "Other natural vegetation" includes grasslands, savannas, and shrublands. Error bars show 95% confidence intervals. TWS = terrestrial water storage.

Addressing distortions in agricultural water pricing and strengthening integrated water resources management can help preserve freshwater resources.

FIGURE O.3 Impact of pricing and the degree of IWRM on freshwater reserves

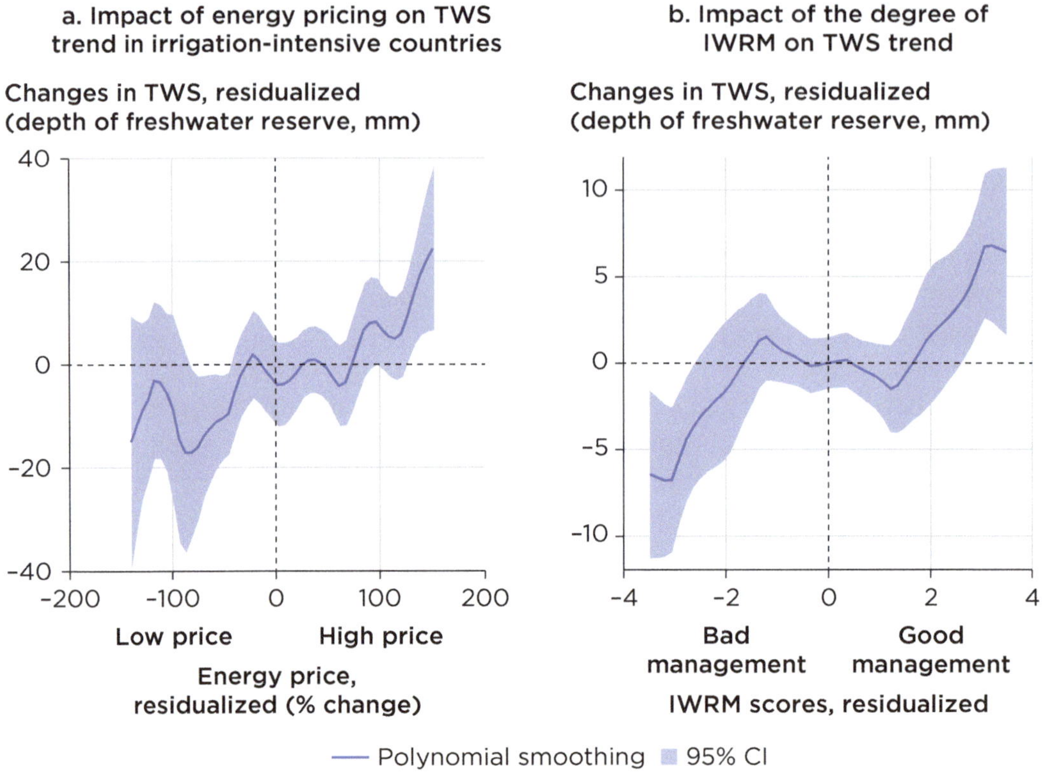

Source: World Bank.

Notes: Panel a illustrates the relationship between changes in energy prices and changes in freshwater reserves (measured by TWS) for countries where more than 20 percent of cropland is irrigated. Panel b illustrates the relationship between the score of IWRM and changes in TWS. All variables are adjusted (residualized) to account for yearly weather differences and country-specific factors. Refer to the online technical appendixes for details. The unit of the *y* axis—mm depth—refers to the equivalent depth of TWS in millimeters, that is, the depth of water when it is uniformly distributed across each 0.5-degree grid cell. CI = confidence interval; IWRM = integrated water resources management; TWS = terrestrial water shortage.

The cascading impacts of continental drying

Continental drying means that water is becoming scarcer in large parts of the world. A growing body of literature has shown that water scarcity reduces agricultural productivity, slows economic growth, affects health and education, and triggers forced migration and potential conflicts. Additionally, water storage, such as groundwater and soil moisture, ensures water availability when precipitation and surface water supplies decline during droughts. The large-scale decline in groundwater and soil moisture weakens this crucial insurance mechanism against climate variability.

Building on previous studies, this report delves into areas that have not been fully explored in the existing literature. The findings highlight the far-reaching, cascading impacts of continental drying on jobs, income, and the environment. Specifically, the analysis yields the following four main takeaways:

1. Freshwater depletion is jeopardizing jobs and incomes.
2. Adaptation has been limited, and the negative effects of warming and drying on food security reinforce one another, risking a potential tipping point.
3. Local water shortages can have large spillover effects globally because of interconnected trade networks.
4. Continental drying intensifies the frequency and severity of wildfires and poses a serious threat to biodiversity.

From wilted fields to waning work

Water availability affects labor market outcomes through both supply- and demand-side channels. On the supply side, water scarcity affects labor productivity by harming nutrition and health and long-term human capital development. During water shortages, immediate productivity losses can occur if people fall ill or spend productive hours fetching water or caring for those who are sick. Illness can result from reduced access to clean water, increasing the risk of waterborne diseases, and from malnutrition caused by droughts.

On the demand side, water shortages negatively affect productivity in water-dependent sectors such as agriculture, energy, manufacturing, tourism, and transportation. This decline in productivity lowers job demand and negatively affects income. About 78 percent of jobs in the global workforce have some level of water dependence, and 42 percent of jobs are significantly water dependent (Connor and Chaves Pacheco 2024).

Water's role in sustaining jobs and livelihoods is even more pronounced in developing countries, where a large share of the population depends on agriculture. About 80 percent of the world's poor live in rural areas and rely heavily on farming, which requires reliable access to water for crop cultivation. Globally, 1.23 billion people are directly employed in agriculture. This dependence is particularly striking in regions such as

South Asia and Sub-Saharan Africa, where more than 50 percent of the workforce relies on agriculture. In these areas, dry shocks can swiftly erode job opportunities, particularly in rural farming communities, leaving livelihoods at risk.

In Sub-Saharan Africa, the analysis shows that a *dry shock*—defined as soil moisture proxied by a drought index of 1.5 standard deviations (SD) below the long-term mean—reduces the employment rate by 2.5 percentage points on average and by 7.5 percentage points in agriculture-dependent rural areas, all else being equal. Given the frequency of dry shocks, this effect translates to approximately 600,000–900,000 individuals in Sub-Saharan Africa being jobless each year as a result of exposure to dry shocks between 2005 and 2018. These numbers represent a loss of 7–9 percent of the annual jobs created in the region. Women, older individuals, landless farmers, and low-skilled workers are those most negatively affected (refer to figure O.4).

Droughts most negatively affect jobs for women, older individuals, landless farmers, and low-skilled workers.

FIGURE O.4 Effects of drought on employment, by worker characteristics

Source: Khan et al. 2024.
Note: The figure depicts the marginal effect of a 1.5-standard-deviation decline in the Standardized Precipitation-Evapotranspiration Index on the probability of having a job across demographic groups. The error bars represent the 95 percent confidence interval.

Of course, job outcomes are not just about numbers—the types of jobs matter, too. Although water scarcity initially reduces agricultural employment, its long-term effect is more complex. In response to prolonged droughts, local populations may convert more land to farming to compensate for lost productivity (Damania et al. 2017), potentially increasing agricultural employment. In low-income countries facing trade barriers, shrinking agricultural productivity may push workers further into subsistence farming, reinforcing dependence on agriculture (Nath 2024).

An increase in agricultural jobs in water-stressed economies is not necessarily a positive outcome. Water scarcity reduces agricultural productivity, leading to lower real income for farmers. Moreover, heavily agriculture-dependent societies in water-stressed regions remain highly vulnerable to future water crises. Economies that have achieved greater economic diversification with more occupations in services and industries are better equipped to withstand dry shocks.

Drying, warming, and a tipping point

This report reveals that the adverse effects of drying and warming are mutually reinforcing. Drying makes adapting to warming more difficult, and warming exacerbates the negative effects of drying. Moreover, these negative effects are nonlinear, implying that the cumulative effects of drying and warming may lead to a tipping point at which their joint impacts on food production are exponentially greater.

Specifically, figure O.5, panel a, illustrates that in warmer regions, as baseline precipitation decreases, even a small decrease in precipitation leads to exponentially greater losses in crop yields. Similarly, figure O.5, panel b, shows that, in the drier half of the world, the adverse effects of warming on crop yields double in regions when baseline temperatures exceed 20°C and quintuple when they exceed 25°C.

The joint effects of drying and warming are potentially staggering. In some of the world's hottest and driest regions, a 1 SD decline in soil moisture proxied by the drought index can lead to a 43 percent reduction in crop productivity. The report finds that drought-prone regions are particularly vulnerable to declining water availability, revealing the limited effectiveness of current adaptation efforts.

Negative effects of drying and warming on crop yields are mutually reinforcing.

FIGURE O.5 Marginal effects of drying and warming on vegetation yields, by climate zone

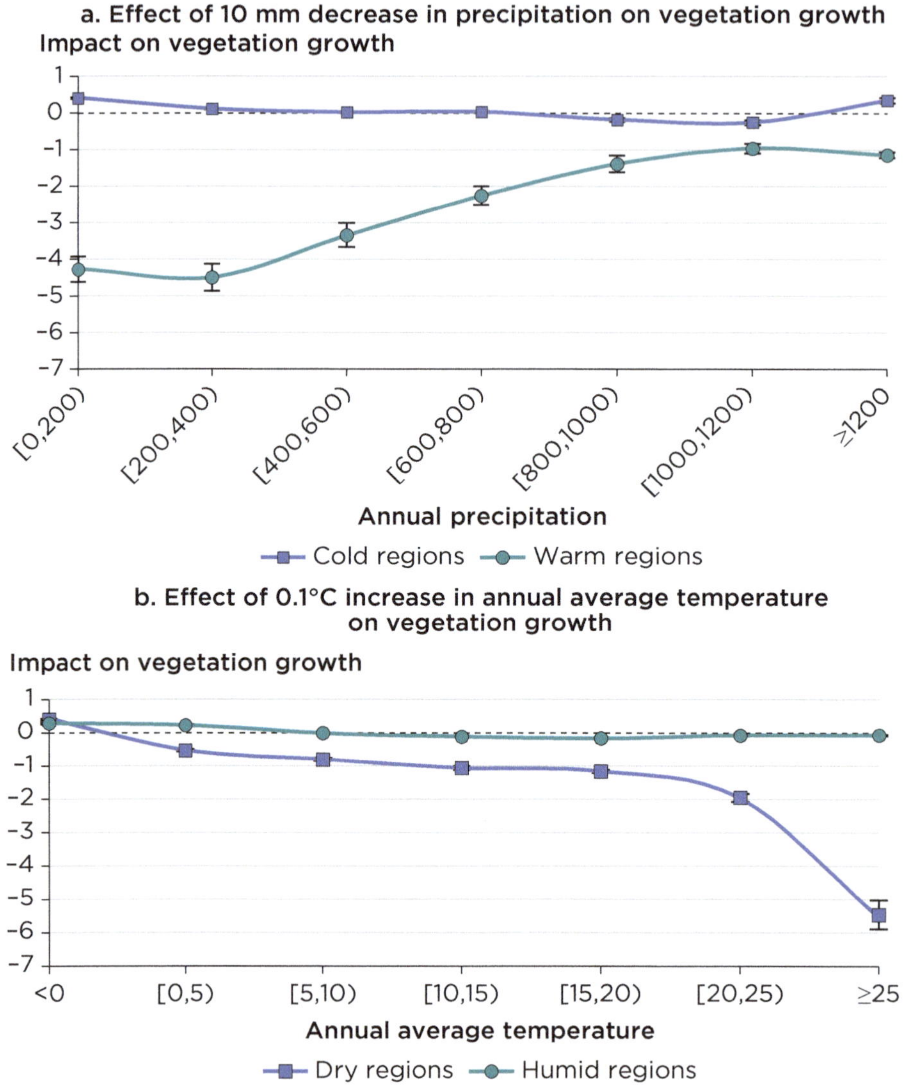

Source: World Bank.

Note: Panel a presents the average marginal effects of a 10 mm decrease in annual total precipitation below the long-term average on (log) NPP—a proxy for vegetation output—for cells within each climate zone. Panel b depicts the average marginal effects of a 0.1°C increase in annual average temperature above the long-term average on (log) NPP for cells within each climate zone. Climate zones are defined by the long-run average temperature and precipitation of each cell over the period 1951–2022, categorized using 5°C bins for temperature and 200 mm for precipitation. Climate bins are defined with inclusive lower bounds and exclusive upper bounds (e.g., [5–10) includes 5 but not 10). The global long-term median temperature (16.8°C) serves as the threshold between warm and cold regions in panel a. The global long-run annual median precipitation (521 mm) serves as the threshold between dry and humid regions in panel b. NPP = net primary product.

Local water shocks, global economic impact

GCEW (2024) highlights the importance of valuing and governing the hydrological cycle as a global common good because of its interconnected nature. This report highlights that increased agricultural trade also heightens water dependencies across countries.

The removal of tariff and nontariff trade barriers has driven significant growth in agricultural trade over recent decades (OECD 2019; OECD and FAO 2021). Between 2018 and 2020, the value of agricultural trade in crops increased by more than 80 percent compared with the 2007 value, whereas manufacturing exports rose by less than 40 percent. Consequently, the water embedded in the production of goods and services traded globally, known as *virtual water trade,* increased by 26 percent between 2000 and 2019. In 2019, approximately 25 percent of global water consumption was allocated for export rather than domestic use.

Given the large scale of virtual water trade, local water shortages can have far-reaching global consequences. The report's simulation analysis illustrates both the magnitude and the channels through which agricultural productivity losses due to water scarcity in one country can ripple across sectors and national borders. For example, a 100 mm drop in annual rainfall in India—roughly two-thirds of a standard deviation from the annual average—is estimated to reduce global real income by $68 billion. Although impacts are primarily felt in the agricultural sector, they also extend to the manufacturing and service sectors that rely on agricultural inputs, such as packaged food production and the restaurant industry.

Trade can expose countries to the effects of water shortages even if they are not directly experiencing drying, but it also plays a crucial role in mitigating these impacts. This report finds that, when consumers shift their spending in response to water shocks and source food and water-intensive goods from unaffected regions, trade helps cushion the local adverse effects of dry shocks, stabilizing prices and sustaining consumption. Countries can improve their resilience to climate-related shocks by establishing strategic trade partnerships with regions that are less susceptible to similar climatic risks.

Parched landscapes, intensified wildfires

Water is essential for sustaining life and maintaining a livable planet. This report examines the effects of continental drying on ecosystems and biodiversity, particularly through its influence on wildfires. Although wildfires are often naturally occurring phenomena, they can be catastrophic for ecosystems and nearby human communities.

This report finds that continental drying significantly increases both the frequency and the magnitude of wildfires, especially in biodiversity hot spots. A 1 SD increase in the rate of TWS depletion raises the probability of wildfire occurrence in biodiversity hot spots by 50 percent, compared with 11 percent outside these areas. On average, a 1 SD increase in the rate of TWS loss leads to a 27 percent higher chance of wildfire occurrence and a 46 percent increase in the size of burned areas, all else being equal. At least 17 of the 36 globally recognized hot spots are experiencing sustained reductions in freshwater availability (refer to map O.2).

Freshwater decline has the most significant impact on wildfires in areas with the lowest gross domestic product per capita and local adaptive capacity, as proxied by the Human Development Index. The effect in these regions is 6 to 9 times greater than in the next quartile, regardless of land cover or climate zones. These findings indicate that, although continental drying leads to greater risk of wildfires, socioeconomic development—including factors affecting human behaviors, wildfire

Almost half of global biodiversity hot spots are experiencing persistent drying.

MAP O.2 Trends in TWS across global biodiversity hot spots

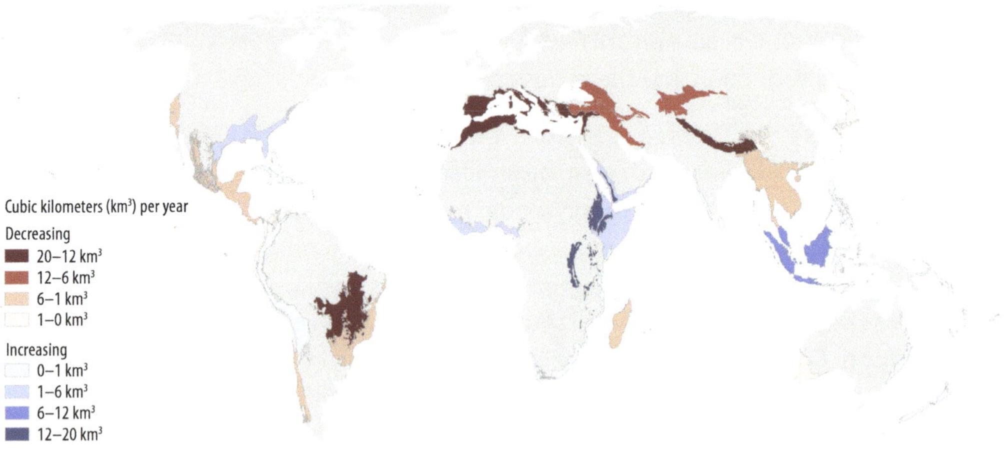

Source: World Bank.
Note: The map plots the annual rate of TWS change in biodiversity hot spots. Red denotes decreasing TWS; blue denotes increasing TWS. Notably, some hot spots in Southeast Asia and Sub-Saharan Africa display increasing TWS. However, this trend may indicate intensified flooding or human interventions, such as the development of large-scale dams, which can also devastate biodiversity. TWS = terrestrial water storage.

prevention measures, and community practices—can play a critical role in mitigating these risks.

Water vulnerability hot spots and priority regions

Water consumption patterns determine a region's water dependency and long-term resilience to water scarcity. Regions with rapidly rising water consumption, low water use efficiency, and more water-intensive economic activities are more susceptible to the impact of continental drying. Agriculture is the largest water consumer, accounting for 98 percent of the global water footprint. By combining water availability and agricultural water demand data, this report identifies vulnerability hot spots and priority regions for policy interventions.

Vulnerability hot spots

This report shows that global water consumption increased by 25 percent between 2000 and 2019, with about a third of this increase occurring in drying regions. These regions include areas already facing freshwater scarcity, such as Central America, northern China, a large swath of Eastern Europe, northern India, and the southwestern United States. However, water stress is also emerging in historically water-abundant regions undergoing rapid agricultural, industrial, and urban growth, such as southeastern Brazil.

Adding to the pressure, a significant share of crop production in drying regions is inefficient, meaning water use per ton of crop produced exceeds that of at least half of global producers under similar climatic conditions and production systems (irrigation or rain fed). Inefficient water use may have contributed to drying in the first place and continue driving water depletion of already-stressed water systems as producers extract more water to compensate for inefficiencies.

Notably, about one-quarter of inefficient water consumption in rain-fed agriculture and one-third in irrigated agriculture are concentrated in regions experiencing declining freshwater availability. These hot spots where water inefficiency coincides with drying trends are most pronounced in Western Asia, Eastern Europe, and North Africa (refer to map O.3). On a national level, the highest share of inefficient agricultural water consumption under drying conditions is observed in Algeria, Cambodia, Mexico, Pakistan, Thailand, Tunisia, and Romania.

A large share of agriculture production in drying regions relies on inefficient water use.

MAP O.3 Share of agricultural water use with low efficiency in drying regions

a. Rain-fed areas

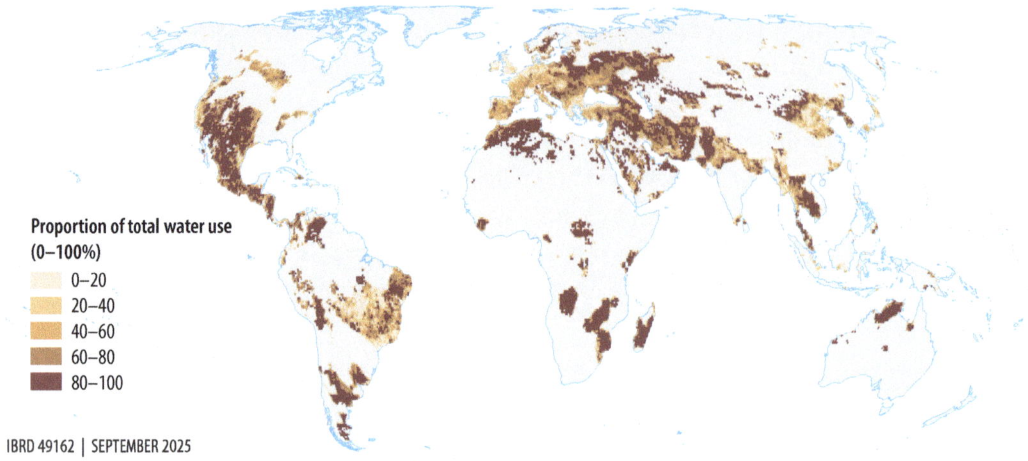

b. Irrigated areas

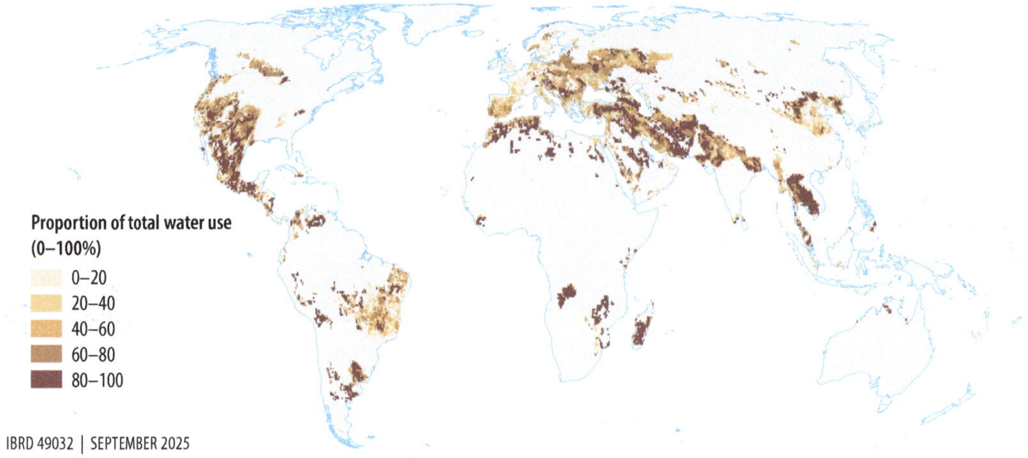

Source: World Bank.

Note: Percentages indicate the share of water use of 35 key crops produced with low water use efficiency in rain-fed and irrigated production systems in drying regions. *Low water use efficiency* is defined as consuming more water per ton of output than at least half of global producers using the same technology and within the same climate zone.

Beyond water use efficiency, crop choice plays a key role in determining overall water demand. The past two decades have seen a global shift toward the cultivation of more water-intensive crops. Among drying countries, 37 have transitioned to more water-intensive agriculture, including 22 located in arid and semiarid regions. This structural shift, coupled with inefficiency, further intensifies water demand in already water-stressed countries. More than two-thirds of the inefficient irrigation in drying areas is linked to the cultivation of water-intensive crops, such as rice, wheat, cotton, maize, or sugar cane.

Furthermore, many water-scarce countries engage in inefficient and unsustainable virtual water exports, using large volumes of water to grow water-intensive crops for export rather than domestic use (refer to map O.4). Inefficient virtual water trade occurs when exporting countries use more water than importers to produce the same crops. Additionally, if the exporter relies more heavily on water from arid or water-stressed regions than the importer does, the exchange is considered unsustainable for the exporter. Between 2000 and 2019, about 17 percent of virtual water export from drying and arid countries was deemed *suboptimal*—defined as inefficient and unsustainable—compared with 11 percent of such exports from nondrying countries.[2] The share of suboptimal water trade embedded in irrigated crops from drying and arid countries was even higher, reaching 27 percent over the same period.

Water-saving potential for drying regions

Improving crop water use efficiency has the potential to save significant amounts of water. This report estimates that, if all low-efficiency producers in drying regions improved to at least global median levels of water use efficiency, irrigation water consumption for 35 key crops—which accounts for more than 90 percent of global irrigation water use—could be reduced by 18 percent. The largest potential for irrigation water savings is found in South Asia. Globally, aligning the production of the 35 key crops with global median levels of water use efficiency could reduce annual irrigation water consumption by 137 billion cubic meters—equivalent to the annual water needs of 118 million people.

Note that efficiency improvement alone does not guarantee water savings. Without proper regulation, efficiency gains may encourage expanded agricultural activity or the adoption of more water-intensive crops, ultimately intensifying water use and exacerbating scarcity—a phenomenon known as *Jevon's Paradox* (Grafton et al. 2018). Combining technological advancements with pricing incentives and water use quotas is key to ensuring that improved efficiency translates into genuine water savings.

Many water-scarce countries engage in suboptimal virtual water export.

MAP O.4 Share of suboptimal virtual water exports embedded in irrigated crops, by country of origin

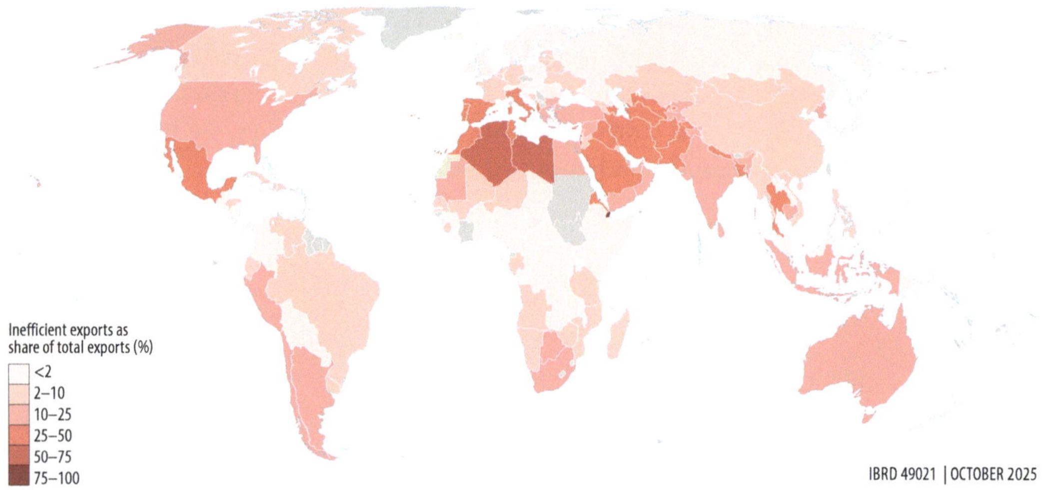

Source: World Bank.
Note: Countries shaded in gray export little to no virtual water. Increasing color intensity indicates increasing share of suboptimal virtual water export embedded in irrigated crops as a share of the total. Virtual water export is considered suboptimal when (1) exporting countries use more water than the importer to produce the same crops, indicating no net water savings, and (2) the exporter relies more heavily on water from arid or water-stressed regions compared to the importer, which suggests the exchange may be less sustainable for the exporter.

Another strategy for water savings is to improve spatial efficiency, including adjusting cropland distribution to better align with water availability within national borders or reallocating water use from less efficient to more efficient producers and from water-scarce to water-abundant areas across countries through virtual water trade. By leveraging differences in water use efficiency among countries, this report estimates that virtual water trade led to annual global water savings of 475 billion cubic meters per year, equivalent to 9.4 percent of total water consumption of the 35 key crops between 2000 and 2019. However, on average, drying countries have gained less from virtual water trade than their nondrying counterparts. Although virtual water trade contributes to annual water savings equivalent to 11 percent of annual water consumption in nondrying countries, it contributes only 8 percent in drying countries.

Trade outcomes are influenced by various factors of comparative advantage, such as capital and labor. Studies have shown that water also plays a crucial role, especially in water-intensive sectors (Carleton, Crews, and Nath 2025; Debaere 2014; Lai, Li, and Zhang forthcoming). However, distortions in agricultural input and output markets—such as energy subsidies for groundwater pumping and above-market price guarantees for certain crops—can reinforce comparative disadvantage. These policies may hinder structural transformation and lead water-scarce countries to continue producing and exporting water-intensive products, further straining their already limited resources. Although agricultural production must balance multiple objectives, such as food security, economic growth, jobs creation, and rural development, prioritizing water efficiency and savings by removing pricing distortions can lead to more sustainable outcomes.

Priority regions

Countries vary widely in the severity and nature of their water challenges; some are driven by supply-side constraints, others by rising demand, and many by a combination of both. To help identify priority regions, two composite indexes are developed to reflect supply and demand stressors at the country level. The supply-side index measures the rate of water depletion and baseline water endowment (proxied by aridity). The demand-side index measures efficiency in crop water use, the water intensity of crops, and the share of suboptimal virtual water export. The placement of each region corresponds to the population-weighted mean of the index across all countries within the region (refer to figure O.6).

Regional averages can mask significant variations at the country level. For example, in Sub-Saharan Africa, countries such as the Democratic Republic of Congo do not face physical water supply constraints, whereas countries such as Chad, Mauritania, Niger, and the Federal Republic of Somalia do. Moreover, in large countries such as China, Mexico, and the United States, national averages may overlook important subnational differences. For example, northern China and the southwestern United States are already facing severe water stress and are part of the mega-drying regions identified in map O.1.

Priority regions can be identified based on supply- and demand-side stressors.

FIGURE O.6 **Composite indexes measuring supply- and demand-side stressors**

a. Supply: Drying and arid (low to high)

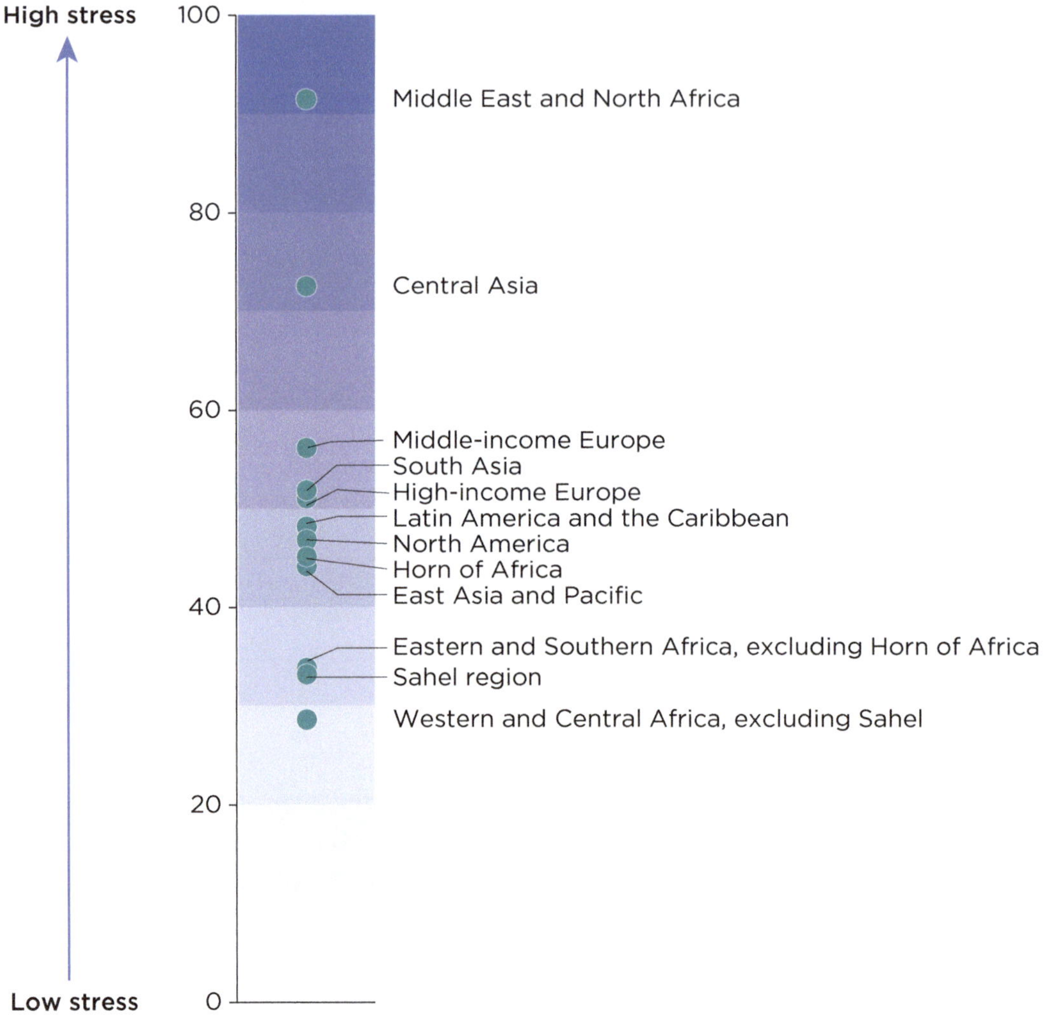

(continued)

FIGURE O.6 Composite indexes measuring supply- and demand-side stressors *(continued)*

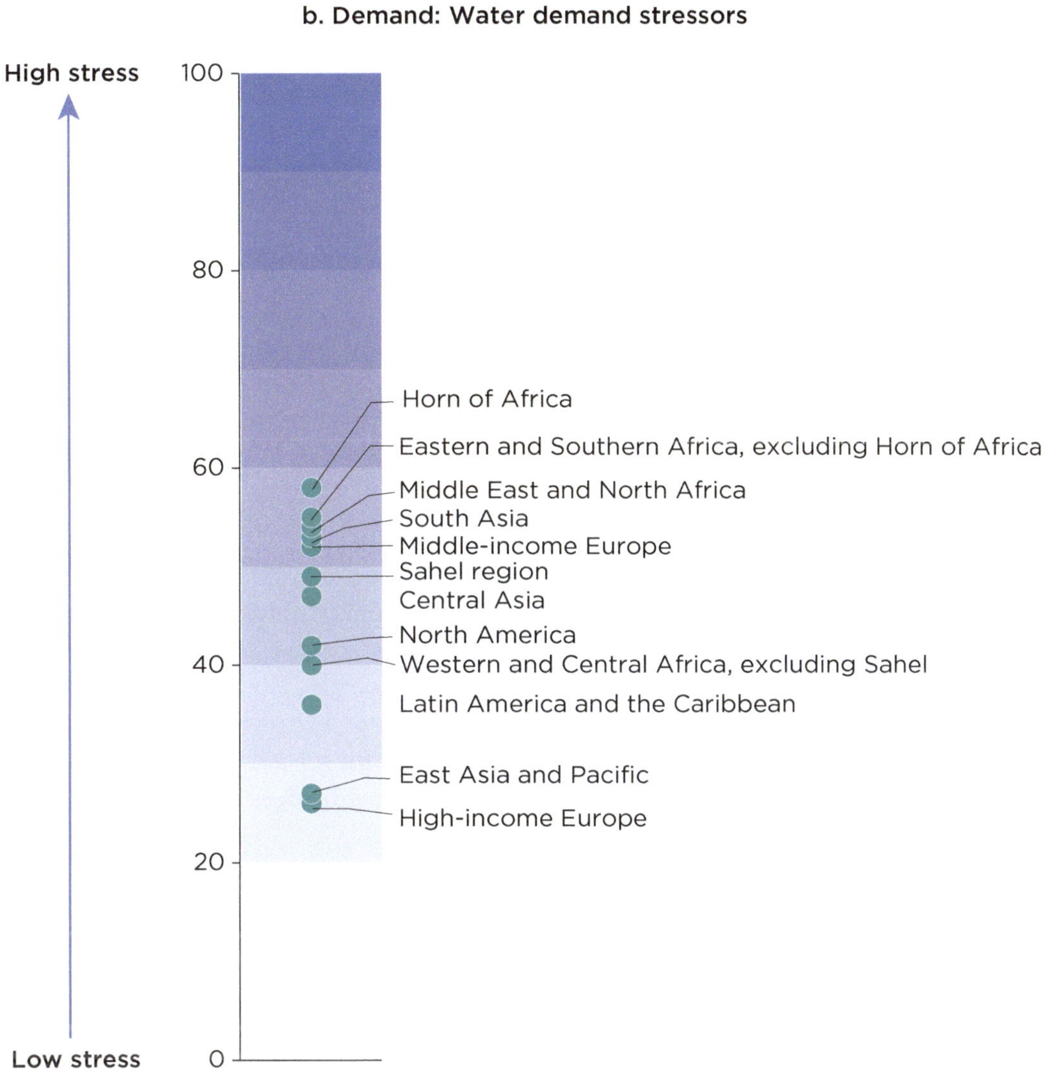

Source: World Bank.

Notes: The supply index considers how fast countries are drying as well as their aridity. The demand index focuses on crops and considers efficiency and water intensity in production as well as efficiency in virtual water trade. The placement of each region corresponds to the population-weighted mean of the index across all countries within the region.

Policy recommendations

This report recommends a three-pronged strategy to address the continental drying crisis through targeted actions in the water sector:

1. Managing water demand by increasing water use efficiency, setting extraction limits, and raising public awareness;
2. Augmenting water supply through water recycling, desalination, and improved natural and built storage; and
3. Improving water allocation to ensure scarce water resources are distributed fairly and efficiently.

Five cross-cutting policy levers are essential for implementing the strategy: strengthening institutions, optimizing water tariffs, adopting water accounting, leveraging data and technology innovations, and valuing water in trade.

Managing water demand

As highlighted in the previous section, improving water use efficiency offers significant potential for water savings. This report outlines three key approaches to managing demand: adopting water-efficient technologies, enforcing extraction limits, and promoting public awareness and education. Water pricing provides necessary incentives for the adoption of water-efficient technologies and is discussed in more detail as a cross-cutting policy lever.

Water-efficient technologies, such as drip irrigation, alternate wetting and drying, and the system of rice intensification, have been demonstrated to reduce agricultural water use by 20 to 50 percent while maintaining crop yields. In addition to these methods, other innovative agricultural practices can contribute significantly to reducing water demand. One approach is adopting drought-resistant crops. Certain crops, such as sorghum and millet, are more drought-resistant than maize and rice and provide similar nutritional benefits. Choosing drought-tolerant varieties of a given crop can enhance water use efficiency and improve yield per water unit. Another approach to reducing water demand is applying deficit irrigation, which induces mild water stress to conserve water with minimal effects on yield. Additionally, adjusting the cropping calendar to align planting schedules with periods of higher rainfall can reduce the need for supplemental irrigation.

As discussed earlier, improving water use efficiency alone does not always lead to water savings. To achieve water conservation, many countries have adopted water extraction limits. These regulations have, in turn, driven the adoption of efficient technologies, reinforcing conservation efforts (Pérez Blanco, Hrast-Essenfelder, and Perry 2020).

The adoption of water-efficient technologies faces several challenges. Farmers often favor traditional practices because of familiarity and risk aversion, and a perceived abundance of water further discourages change. Because continental drying is a slow and largely invisible process, the urgency of conservation often goes unrecognized until a crisis, such as a severe drought, occurs.

To overcome these barriers, public awareness campaigns play a crucial role in driving behavioral change. In the Calamba Water District in the Philippines, efforts to reduce water leakage—also known as *nonrevenue water*—through public awareness and engagement encouraged local communities to report leaks and misuse, fostering a culture of conservation that complemented infrastructure improvement initiatives (Cervancia et al. 2022). Additionally, providing proper training and ongoing support to farmers is essential for encouraging the adoption of more efficient water management practices.

Augmenting water supply

Unconventional water, such as recycled and desalinated water, offers new possibilities for expanding freshwater availability. Water recycling transforms used water into a valuable resource by treating and reusing it for agriculture, industry, and even drinking water. Instead of being lost as discharge, reclaimed water becomes a sustainable supply that reduces reliance on traditional sources. Similarly, desalination extracts salt and impurities from seawater or brackish water, converting it into fresh, potable water. By harnessing these technologies, communities can build more resilient water systems in the face of growing demand and declining supply.

Over the past two decades, global installed reuse capacity has increased, yet progress remains modest, with capacity reaching only 183 million cubic meters per day by the end of 2024. Meanwhile, more than 20,000 desalination plants operate across more than 150 countries, providing drinking water to an estimated 300 million people (World Bank 2019). Despite its potential, water recycling faces significant obstacles, including high treatment costs and public resistance, particularly to potable reuse. Even when used for irrigation, recycled water must be carefully managed to prevent soil salinization.

Desalination also faces economic and environmental challenges. The process remains costly, often two to three times more expensive than conventional sources (Ziolkowska 2015), although costs have been declining in recent years, driven by the increased use of low-cost renewable energy. Additionally, desalination generates concentrated brine waste that can harm marine ecosystems if not properly managed (Antia 2022). Inland desalination poses additional risks because brine disposal can have a negative effect on land and groundwater.

Establishing clear quality standards and regulatory frameworks for recycled and desalinated water will be critical to ensure safety and environmental sustainability. To improve economic feasibility, advancements in cost reduction, renewable energy integration, and public acceptance—supported by policy measures and education—will be essential. When properly implemented, recycling programs and desalination can provide a reliable, drought-resistant source of high-quality water.

Another essential aspect of enhancing water availability is the expansion of both natural and built water storage. Groundwater recharge can be facilitated through check dams and percolation tanks, which help capture surface runoff and enable its percolation into the ground, thus recharging aquifers. Conservation agriculture practices, such as crop rotation and mulching, improve water infiltration and retention, thereby conserving soil moisture. Rainwater harvesting, such as building small ponds and water retention structures, captures and stores rainwater for agricultural use. In addition, as highlighted earlier, the conservation of forests and wetlands safeguards natural storage capacities. Furthermore, sustainable sediment management practices are crucial to increase the capacity of built storage.

Improving water allocation

Effective water allocation—determining who gets how much water, when, and for what purpose—is the backbone of sustainable water management. Improving water allocation is more critical than ever in the face of growing competition among agriculture, industry, domestic needs, and ecosystems in a drying world. Effective allocation depends on the ability to reallocate water as supply and demand fluctuate, relying on a mixture of administrative procedures, judicial rulings, water user associations, and market-based mechanisms (Garrick, Chautard, and Rawlins 2019; Marston and Cai 2016; Meinzen-Dick and Ringler 2008).

Water tenure or water rights systems facilitate efficient and equitable water sharing. The effectiveness of water tenure systems depends on clearly defining water rights within the legal framework of water allocation; establishing and regulating the legal and informal mechanisms that allow the transfer of water rights or access among individuals, communities, or institutions; and adopting governance structures focusing on enforcing payments and fees, regulating the final volumes of water transferred, and managing dispute resolution processes. Strong water legislation, enforceability, and reallocation mechanisms are critical elements of effective water tenure systems.

Incorporating environmental water flows into water allocation decisions is essential to achieve long-term sustainability and resilience in the water sector. Freshwater ecosystems rely on a flow regime (including baseflows and flooding to maintain the functions and services delivered by rivers, wetlands, and aquifers, such as water purification, flood regulation, biodiversity support, and climate resilience). These freshwater ecosystem services directly and indirectly sustain human well-being and economic activity. Recognizing environmental water flows as a legitimate and necessary use is a core component of IWRM. Several countries have begun to embed this principle in their water governance frameworks. In Chile, the 2022 Water Code Reform marked a significant step forward by introducing a requirement to maintain minimum environmental flows.

Transboundary cooperation is key to fair and sustainable water sharing across and within basins. Strong frameworks—such as joint management bodies and basin-level agreements—help riparian countries work together, improve resource management, and reduce the risk of conflict. Basin transfer schemes within national boundaries are also feasible mechanisms to transfer water from wet to dry areas.

Finally, integrating virtual water trade into water allocation strategies helps mitigate water shortages in water-scarce regions. At the global level, more strategically harnessing virtual water trade—linking water-rich regions with water-scarce ones—could relieve pressure on depleted resources and support fast-growing economies (Flach et al. 2016; Vallino, Ridolfi, and Laio 2021).

Five cross-cutting levers

Five cross-cutting levers are critical for advancing water supply, demand, and allocation solutions. Table O.1 outlines a phased policy action road map for implementing the proposed policy levers.

TABLE O.1 Policy roadmap for the water sector to address the continental drying crisis

	Policy action		
Policy lever	**Short term (0–1 year)**	**Medium term (1–3 years)**	**Long term (>3 years)**
Strengthen institutions	• Identify institutional gaps in water allocation and monitoring. • Launch capacity-building programs for local water agencies. • Develop national water efficiency standards. • Undertake a rapid assessment of environmental water requirements for critical ecosystems.	• Establish interagency coordination mechanisms for IWRM. • Establish regulatory frameworks for groundwater management, including licensing and abstraction limits. • Strengthen decentralized water governance. • Establish clear PPP frameworks to promote private investment in water.	• Promote transboundary water cooperation mechanisms. • Institutionalize adaptive water planning processes informed by data, climate scenarios, and stakeholder engagement.
Reform water tariffs and repurpose subsidies	• Conduct water tariff and subsidy reviews, including assessing implicit subsidies (such as unmetered water, subsidized energy for groundwater) to inform the design of cost-recovery tariffs. • Evaluate the distributional impacts of tariff reforms and assess whether existing social safety nets are sufficient to ensure affordability.	• Launch a public communications campaign to build support for reform. • Pilot (a) volumetric tariffs for agricultural water supply, (b) performance-based tariffs, and (c) repurposing of agricultural water subsidies to support water-saving practices (such as drip irrigation) and payment for environmental services for forest and wetland protection in selected high-use areas. • Enhance the social safety net as necessary to ensure affordability.	• Scale up cost-recovery tariffs. • Expand performance-based tariffs and subsidies linked with conservation outcomes. • Establish a national authority to continuously monitor water pricing and implement cost-recovery tariffs and targeted subsidies.
Adopt water accounting	• Conduct an audit of water use, starting with synthesizing existing data on water consumption and losses. • Determine the amounts of water formally allocated for consumptive and nonconsumptive uses, to identify potential reserves or resources available for environmental flows.	• Launch public data portals that visualize water flows and sectoral usage to build public awareness and accountability. • Develop water-efficiency benchmarks across basins and sectors using water accounting.	• Institutionalize water accounting to guide water resources planning and allocation.

(continued)

TABLE O.1 Policy roadmap for the water sector to address the continental drying crisis *(continued)*

	Policy action		
Policy lever	**Short term (0–1 year)**	**Medium term (1–3 years)**	**Long term (>3 years)**
Leverage data and technology innovation	• Deploy smart meters to enable volumetric pricing and detect overuse. • Pilot the use of downscaled GRACE data for near-real-time monitoring of water storage change. • Introduce AI-driven systems to enable detection of leaks and optimization of irrigation scheduling. • Introduce Internet of Things technologies (for example, soil moisture sensors and mobile platforms) to support smallholder farmers' decision-making.	• Institutionalize satellite-based tracking of water storage change (for example, using downscaled GRACE). • Expand decentralized water decision-making with AI-powered digital controls and sensors.	• Promote in situ water monitoring and data sharing to further enhance the resolution of GRACE data. • Scale precision agriculture using AI-driven irrigation advisory systems, drone spraying, and cloud-based water scheduling. • Integrate real-time allocation models with legal water rights systems, ensuring enforcement and adaptive reallocation during droughts.
Value water in trade	• Quantify the water footprint of key agricultural and industrial exports, especially in water-stressed basins. • Promote public awareness of the water footprint of imports by, for example, creating water intensity labels for imports.	• Establish interministerial platforms (for example, water-agriculture-trade) to coordinate planning and ensure that water is a factor in strategic trade decisions. • Explore technical quotas on virtual water imports based on the sustainability of trade partners' water management.	• Integrate virtual water balances into national food and trade security strategies, identifying optimal import-export mixes to promote water and food security. • Institutionalize water valuation tools in trade negotiations and agreements, particularly in regions reliant on export agriculture (for example, floriculture, rice, or cotton).

Source: World Bank.
Note: AI = artificial intelligence; GRACE = Gravity Recovery and Climate Experiment; IWRM = integrated water resources management; PPP = public-private partnership.

Strengthening institutions

This effort involves the establishment and enforcement of institutional frameworks and formal rules that promote transparent and equitable resource distribution. The effectiveness of water pricing mechanisms depends on robust monitoring and enforcement to ensure compliance and prevent illegal extraction, especially in water-scarce regions (Grafton et al. 2018). Effective IWRM depends on institutions working across sectors—agriculture, forestry, water, and urban planning—to transform land and water management practices. Successful transboundary water cooperation requires strong coordination among multisectoral governance stakeholders.

Reforming water tariffs and repurposing subsidies

As the analysis in this report shows, undervaluing and underpricing of water encourage excessive use and inefficiency, exacerbating continental drying. Cost-reflective water tariffs are essential to provide incentives for adopting water-saving practices and technologies. They are also critical for attracting supply-side investment, including from the private sector, and the resulting revenue can be used to fund water infrastructure or broader water-saving initiatives. Water pricing and tradable water rights help reallocate water efficiently.

Proper compensation mechanisms are crucial to protect vulnerable populations—such as smallholder farmers and low-income households—from negative financial burdens. Studies suggest that water pricing faces less political challenge when the revenue generated is reinvested in social safety nets or alternative livelihood programs to mitigate adverse impacts (Jia et al. 2023). Moreover, repurposing subsidies can yield a double dividend. For instance, redirecting water and energy subsidies to targeted support for drip irrigation and incentivizing farmers to protect wetlands, forests, and groundwater recharge—through payments for environmental services—can simultaneously reduce water waste and promote sustainable resource management. Performance-based tariffs that link water charges to consumption levels and service performance can create incentives for conservation.

Adopting water accounting

Water accounting—that is, accurate measurement of water consumption and supply—is essential to design tariffs that reflect the true cost of water services and to effectively target subsidies. Water accounting also supports IWRM and water tenure systems. Accurate and timely data on water availability, usage patterns, and flows are essential to inform equitable allocation, detect overuse, and guide adaptive management. Without reliable monitoring, tenure systems risk becoming symbolic rather than operational.

Leveraging data and technology innovations

Analysis of large satellite data sets and artificial intelligence are revolutionizing hydrological assessments and drought monitoring, enabling more precise and efficient water management. As illustrated in this report, integrating downscaled GRACE data with socioeconomic data can help decision-makers to identify unsustainable land and water uses that drive water depletion and to pinpoint hot spot regions where water demand is rising, supply is falling, and efficiency is low. Paired with in situ weather and water monitoring, GRACE data can be further downscaled to hyperlocal levels (approximately 1 km) to enable precision agriculture, boost productivity, and improve livelihoods.

Smart water systems powered by artificial intelligence use real-time data and Internet of Things technologies to monitor water quality, manage irrigation systems, and detect sources of inefficiency. Open data innovations such as remote sensing and evapotranspiration tracking are pushing the envelope even further. In the United States, localized evapotranspiration data are empowering farmers and resource managers to fine-tune water use with unprecedented accuracy (DeMarco 2024). Embracing these technological advancements will be key to enhancing water resilience and ensuring sustainable resource management.

Valuing water in trade

Although global virtual water trade contributes to overall water savings, a significant portion remains inefficient and unsustainable. By incorporating water valuation and recognizing water footprints within trade value chains, countries can optimize virtual water trade. Often, a combination of measures—such as sustainability standards on the supply side and consumer information on the demand side—serves as a mechanism to curb unsustainable water use.

Supply-side measures include technical quotas for trade, which implement trade restrictions when products do not comply with sustainability standards. For example, some countries enforce maximum acceptable limits on pesticide use and agricultural waste to protect land and soil quality. On the demand side, consumer information about the environmental impact of water use and certification and labeling schemes for water-efficient products can help consumers make decisions based on water footprint and sustainability.

Supply- and demand-side interventions, such as expanding water reuse, developing desalination capacity, and scaling up efficient irrigation technologies, often require large capital investment. However, the water sector faces a significant financing gap, and private sector participation is crucial in addressing this gap. The successful implementation of the cross-cutting measures outlined in this report will be critical to attracting private investment. In addition, countries need to establish clear public-private partnership frameworks, adopt a sophisticated mix of financial instruments—such as blended finance and green and blue bonds—and guarantee to reduce investment risk and mobilize private capital at scale. In turn, the private sector can play a vital role in driving innovation and delivering cost-effective, scalable solutions.

Beyond water

Beyond water-specific interventions, economic diversification offers a powerful strategy to enhance resilience and reduce the adverse effects of continental drying on jobs and income. This report shows that communities heavily reliant on water-intensive sectors, such as agriculture, are more vulnerable to drought-related disruptions. Diversification into less water-dependent sectors can help mitigate these risks and build more adaptive local economies.

This finding aligns with evidence in the literature showing that households in regions with a more developed manufacturing sector are better able to offset farm losses from long-term water shortages by increasing off-farm income (Blakeslee, Fishman, and Srinivasan 2020). Additionally, investing in human capital and enhancing market connectivity enable communities to adapt to changing labor market demands and support the growth of a rural nonfarm economy, enhancing income resilience for rural households (Musungu, Kubik, and Qaim 2024; Nguyen, Nguyen, and Grote 2023).

Evidence in this report also suggests that removing trade barriers and establishing strategic trade relationships with countries that are more water abundant and less exposed to similar climate risks can enhance resilience. Such measures not only support local water savings but also help to both manage short-term employment disruptions and address longer-term structural impacts.

Finally, and alarmingly, this report finds that adaptation efforts in agriculture have not been effective at the global level. Key barriers include restricted access to capital, small farm sizes, insecure land tenure, lack of infrastructure, weak institutional support, and risk aversion

(Adger et al. 2009; Below et al. 2012; Core Writing Team, Lee, and Romero 2022; Hornbeck 2012; Morton 2007). There is an urgent need to overcome these barriers through capital and land market reforms to strengthen adaptation and build resilience.

Conclusion

Continental drying—the long-term reduction in freshwater availability across large landmasses—has become an alarming trend and is a growing threat to food, people, and the planet. The solutions, however, are within the reach of most countries, whether poor or rich. They include a three-pronged strategy of managing demand, augmenting water supply, and optimizing water allocation, along with broader measures outside the water sector to strengthen economic resilience and adaptability. Implementing this strategy requires collective action from governments, the private sector, agricultural communities, and water consumers, among others. The challenge is immense, and urgent collaboration is essential to chart a path toward resilience.

Notes

1. *Annual renewable freshwater supply* is the average amount of water newly available for human use every year. This amount includes what remains of rain or snow (precipitation) after accounting for the water lost to the atmosphere through evaporation and plant processes (transpiration) and the water needed to maintain healthy ecosystems and continuation of the water cycle (environmental flows).
2. Some water-scarce countries, such as Israel and Morocco, use desalinated water to produce and export agricultural products. However, available data do not allow for estimation of the share of desalinated versus freshwater used in these virtual water exports or the opportunity cost of this practice, particularly given that desalinated water could be allocated to other sectors.

References

Adger, W. Neil, Suraje Dessai, Marisa Goulden, Mike Hulme, Irene Lorenzoni, Donald R. Nelson, Lars Otto Naess, Johanna Wolf, and Anita Wreford. 2009. "Are There Social Limits to Adaptation to Climate Change?" *Climatic Change 93*: 335–54. https://doi.org/10.1007/s10584-008-9520-z.

Antia, David J. 2022. "Catalytic Partial Desalination of Saline Water." *Water* 14 (18): 2893. https://doi.org/10.3390/w14182893.

Below, T. B., K. D. Mutabazi, D. Kirschke, C. Franke, S. Sieber, R. Siebert, and K. Tscherning. 2012. "Can Farmers' Adaptation to Climate Change Be Explained by Socio-Economic Household-Level Variables?" *Global Environmental Change 22* (1): 223–5.

Blakeslee, D., R. Fishman, and V. Srinivasan. 2020. "Way Down in the Hole: Adaptation to Long-Term Water Loss in Rural India." *American Economic Review* 110: 200–24.

Bryan, Elizabeth, Claire Chase, and Mik Schulte. 2019. *Nutrition-Sensitive Irrigation and Water Management*. Washington, DC: World Bank. https://hdl.handle.net/10986/32309.

Carleton, Tamma, Levi Crews, and Ishan Nath. 2025. "Agriculture, Trade, and the Spatial Efficiency of Global Water Use." Unpublished manuscript, last modified April 11, 2025. https://www.levicrews.com/files/p-wateruse_paper.pdf.

Cervancia, K., C. J. Gomez, C. E. Niega, and K. M. Tividad. 2022. "Assessment, Monitoring, and Reduction Strategy Development for Non-Revenue Water (NRW) of Calamba Water District (CWD), Calamba City, Laguna, Philippines." *IOP Conference Series: Earth and Environmental Science 1022* (1): 012058. https://doi.org/10.1088/1755-1315/1022/1/012058.

Chandanpurkar, Hrishikesh A., James S. Famiglietti, Kaushik Gopalan, David N. Wiese, Yoshihide Wada, Kaoru Kakinuma, John T. Reager, and Fan Zhang. 2025. "Unprecedented Continental Drying, Shrinking Freshwater Availability, and Increasing Land Contributions to Sea Level Rise." *Science Advances*, 11: eadx0298. https://doi.org/10.1126/sciadv.adx0298. Background paper for this report.

Connor, Richard, and S. M. Chaves Pacheco. 2024. *Global Employment Trends and the Water Dependency of Jobs: Technical Paper*. Paris: United Nations Educational, Scientific and Cultural Organization. https://unesdoc.unesco.org/ark:/48223/pf0000388410.

Core Writing Team, Hoesung Lee, and José Romero, eds. 2023. *Climate Change 2023: Synthesis Report*. Geneva: Intergovernmental Panel on Climate Change. https://doi.org/10.59327/IPCC/AR6-9789291691647.

Damania, Richard, Sébastien Desbureaux, Aude-Sophie Rodella, Jason Russ, and Esha Zaveri. 2017. *Uncharted Waters: The New Economics of Water Scarcity and Variability*. Washington, DC: World Bank.

Debaere, P. 2014. "The Global Economics of Water: Is Water a Source of Comparative Advantage?" *American Economic Journal: Applied Economics* 6 (2): 32–48. https://doi.org/10.1257/app.6.2.32.

DeMarco, E. 2024. "OpenET Study Helps Water Managers and Farmers Put NASA Data to Work." NASA. https://science.nasa.gov/science-research/earth-science/water-energy-cycle/openet-moisture-measurement-tool-is-proving-highly-accurate/.

Flach, R., Y. Ran, J. Godar, L. Karlberg, and C. Suavet. 2016. "Towards More Spatially Explicit Assessments of Virtual Water Flows: Linking Local Water Use and Scarcity to Global Demand of Brazilian Farming Commodities." *Environmental Research Letters 11* (7): 075003. https://doi.org/10.1088/1748-9326/11/7/075003.

Garrick, Dustin E., Alice Chautard, and Jonathan Rawlins. 2019. "Reallocating Water." In *Water Science, Policy, and Management: A Global Challenge*, edited by Simon J. Dadson, Dustin E. Garrick, Edmund C. Penning-Rowsell, Jim W. Hall, Rob Hope, and Jocelyne Hughes, pp. 141–58. New York: Wiley.

GCEW (Global Commission on the Economics of Water). 2024. *The Economics of Water: Valuing the Hydrological Cycle as a Global Public Good*. Colombo, Sri Lanka: GCEW, International Water Management Institute.

Grafton, R. Q., J. Williams, C. J. Perry, F. Molle, C. Ringler, P. Steduto, B. Udall, et al. 2018. "The Paradox of Irrigation Efficiency." *Science* 361 (6404): 748–50. https://www.science.org/doi/10.1126/science.aat9314.

Hoekstra, A. Y., and M. M. Mekonnen. 2012. "The Water Footprint of Humanity." *Proceedings of the National Academy of Sciences* 109 (9): 3232–7. https://doi.org/10.1073/pnas.1109936109.

Hornbeck, R. 2012. "The Enduring Impact of the American Dust Bowl: Short- and Long-Run Adjustments to Environmental Catastrophe." *American Economic Review 102* (4): 1477–507.

Immerzeel, W. W., A. F. Lutz, M. Andrade, A. Bahl, H. Biemans, T. Bolch, S. Hyde, et al. 2020. "Importance and Vulnerability of the World's Water Towers." *Nature 577* (7790): 364–9. https://doi.org/10.1038/s41586-019-1822-y.

Jia, Jun-Jun, Tianzhu Lu, Maorong Jiang, and Huaing Wu. 2023. "How to Make Effective Household Water Demand-Side Management Policy in China? Evidence from Survey Data in 50 Cities." *Cities 140*: 104383. https://doi.org/10.1016/j.cities.2023.104383.

Khan, Amjad M., L. Kuate, R. Pongou, and F. Zhang. 2024. "Weather, Water, and Work. Climatic Water Variability and Labor Market Outcomes in Sub-Saharan Africa." Policy Research Working Paper 10823, World Bank, Washington, DC.

Kim, J.-S., K.-W. Seo, T. Jeon, J. Chen, and C. R. Wilson. 2019. "Missing Hydrological Contribution to Sea Level Rise." *Geophysical Research Letters* 46 (21): 12049–55. https://doi.org/10.1029/2019gl085470.

Lai, Wangyan, Shanjun Li, and Fan Zhang. Forthcoming. "The Impact of Water Resources on Trade under a Changing Climate." Background paper, World Bank, Washington, DC.

Marston, L., and X. Cai. 2016. "An Overview of Water Reallocation and the Barriers to Its Implementation." *WIREs Water 3* (5): 658–77. https://doi.org/10.1002/wat2.1159.

Meinzen-Dick, R., and C. Ringler. 2008. "Water Reallocation: Drivers, Challenges, Threats, and Solutions for the Poor." *Journal of Human Development 9* (1): 47–64. https://doi.org/10.1080/14649880701811393.

Morton, J. F. 2007. "The Impact of Climate Change on Smallholder and Subsistence Agriculture." *Proceedings of the National Academy of Sciences* 104 (50): 19680–5.

Musungu, Arnold L., Zaneta Kubik, and Matin Qaim. 2024. "Drought Shocks and Labor Reallocation in Rural Africa: Evidence from Ethiopia." *European Review of Agricultural Economics* 51 (4): 1045–68. https://doi.org/10.1093/erae/jbae020.

Nath, Isan. 2024. "Climate Change, the Food Problem, and the Challenge of Adaptation through Sectoral Reallocation." Working Paper, Federal Reserve Bank of San Francisco, San Francisco, CA.

Nguyen, D. L., T. T. Nguyen, and U. Grote. 2023. "Shocks, Household Consumption, and Livelihood Diversification: A Comparative Evidence from Panel Data in Rural Thailand and Viet Nam." *Economic Change and Restructuring* 56 (5): 3223–55. https://doi.org/10.1007/s10644-022-09400-9.

OECD (Organisation for Economic Co-operation and Development). 2019. *Agricultural Policy Monitoring and Evaluation 2019.* Paris: OECD Publishing. https://doi.org/10.1787/39bfe6f3-en.

OECD (Organisation for Economic Co-operation and Development) and FAO (Food and Agriculture Organization of the United Nations). 2021. *OECD-FAO Agricultural Outlook 2021–2030*. Paris: OECD.

Pérez-Blanco, C. D., A. Hrast-Essenfelder, and C. Perry. 2020. "Irrigation Technology and Water Conservation: A Review of the Theory and Evidence." *Review of Environmental Economics and Policy* 14 (2): 216–39. https://doi.org/10.1093/reep/reaa004.

Reager, J. T., A. S. Gardner, J. S. Famiglietti, D. N. Wiese, A. Eicker, and M.-H. Lo. 2016. "A Decade of Sea Level Rise Slowed by Climate-Driven Hydrology." *Science* 351 (6274): 699–703. https://doi.org/10.1126/science.aad8386.

Rodell, M., A. Barnoud, F. R. Robertson, R. P. Allan, A. Bellas-Manley, M. G. Bosilovich, D. Chambers, et al. 2024. "An Abrupt Decline in Global Terrestrial Water Storage and Its Relationship with Sea Level Change." *Surveys in Geophysics* 45 (6): 1875–1902. https://doi.org/10.1007/s10712-024-09860-w.

Rodell, M., J. S. Famiglietti, D. N. Wiese, J. T. Reager, H. K. Beaudoing, F. W. Landerer, and M.-H. Lo. 2018. "Emerging Trends in Global Freshwater Availability." *Nature* 557 (7707): 651–9. https://doi.org/10.1038/s41586-018-0123-1.

Rodell, M., and B. Li. 2023. "Changing Intensity of Hydroclimatic Extreme Events Revealed by GRACE and GRACE-FO." *Nature Water* 1: 241–8. https://doi.org/10.1038/s44221-023-00040-5.

Scanlon, B. R., A. Rateb, A. Anyamba, S. Kebede, A. M. MacDonald, M. Shamsudduha, J. Small, A. Sun, R. G. Taylor, and H. Xie. 2022. "Linkages between GRACE Water Storage, Hydrologic Extremes, and Climate Teleconnections in Major African Aquifers." *Environmental Research Letters* 17 (1): 014046. https://doi.org/10.1088/1748-9326/ac3bfc.

Vallino, E., L. Ridolfi, and F. Laio. 2021. "Trade of Economically and Physically Scarce Virtual Water in the Global Food Network." *Scientific Reports 11* (1): 22806. https://doi.org/10.1038/s41598-021-01514-w.

World Bank. 2019. "The Role of Desalination in an Increasingly Water-Scarce World." World Bank, Washington, DC. https://hdl.handle.net/10986/31416.

Ziolkowska, J. 2015. "Is Desalination Affordable?—Regional Cost and Price Analysis." *Water Resources Management* 29: 1385–97. https://doi.org/10.1007/s11269.

Glossary

Annual renewable freshwater supply—The average amount of water newly available for human use every year. This amount includes the remainder of rain or snow (precipitation) after accounting for the water lost to the atmosphere through evaporation and plant processes (transpiration) and the water needed to maintain healthy ecosystems and continuation of the water cycle (environmental flows). Polluted water is not typically considered part of a safe or reliable water supply, but it can still be a source of water supply if treated. Because of data constraints, this report does not estimate how the deterioration of water quality affects freshwater supply.

Aridity—A measure of the dryness of a climate, typically defined by the balance between precipitation and potential evaporation. A region is considered arid when it receives significantly less rainfall than the amount of moisture that could be lost through evaporation and plant transpiration, resulting in limited water availability for ecosystems and human use.

Blue water—Usable freshwater from rivers, lakes, and renewable shallow groundwater.

Continental drying—The long-term decline in freshwater reserves across vast landmasses. It refers to both the reduction in total freshwater reserves on land and the emergence of interconnected, continental-scale mega-drying regions.

Freshwater—Nonsalty water found in rivers, lakes, glaciers, and other sources.

Green water—Rainwater stored in the soil and used by plants.

Terrestrial water storage (TWS)—Freshwater stored over land in the form of groundwater, soil moisture, surface water (river, lakes, and reservoirs), snow, glaciers and ice caps, and plants. TWS is a critical indicator of freshwater availability. In this report, the terms *freshwater reserves* and *TWS* are used interchangeably.

Virtual water trade—The hidden flow of water embedded in the production and trade of goods and services. When a country imports products—especially water-intensive ones such as crops, meat, or textiles—it is essentially importing the water used to produce them, rather than using its own water resources.

Water consumption—The amount of water taken that is not returned because of evaporation or absorption into products. For instance, irrigation uses surface or groundwater, but much of it evaporates or is absorbed by crops, so it is considered consumed rather than returned. It is important to distinguish water consumption from water withdrawal.

Water demand—The total quantity of water that an individual or organization requests to meet domestic, industrial, agricultural, and environmental needs within a specific area and time period at a given place. Water demand can be categorized as consumptive demand—water that is used and not returned to its source (for example, irrigation, evaporation)—and nonconsumptive demand—water that is used but largely returned (for example, hydropower, cooling processes).

Water footprint—The quantification of the consumptive use of freshwater, including soil moisture, surface water, and groundwater. The water footprint can measure direct water consumption from the producer's perspective, such as the water used to grow crops. Additionally, it can evaluate indirect water consumption from the consumer's perspective—that is, water consumption in the form of food, energy, and industrial goods and services. A water footprint distinguishes green, blue, and gray components:

- *Green water footprint.* Water consumed from rainwater in the soil that does not become runoff or percolate to groundwater. This water is the most important component of global crop production and the sole source of water in rain-fed production systems.
- *Blue water footprint.* The volume of surface water and groundwater consumed as a result of the production of a good or service.
- *Gray water footprint.* A quantification of water pollution by estimating the volume of freshwater needed to dilute pollutants to meet water quality standards. However, because of limited water pollution data at the required scale and resolution, this report does not include an assessment of the gray water footprint.

Water productivity—Product produced per unit of water consumption. The concept of water productivity is similar to that of labor productivity or land productivity, but with the exception that production is divided by the water input (product units/cubic meters).

Water-stressed area—A (sub)catchment in which the volume of surface water and groundwater consumed surpasses the annual renewable water supply.

Water withdrawal—The total volume of water extracted from surface or groundwater sources for use in agriculture, industry, or households. The withdrawn water will evaporate and return to the original source or to a different catchment or the sea after use. For example, for hydropower, water withdrawn is often returned to the source.

Abbreviations

AI	artificial intelligence
AWD	alternate wetting and drying
CCR	component contribution rate
CDF	cumulative distribution function
CLSM4	Catchment Land Surface Model L4
CSR	Center for Space Research
ECMWF	European Centre for Medium-Range Weather Forecasts
ERA5	ECMWF Reanalysis v5
FAO	Food and Agriculture Organization of the United Nations
FWI	Fire Weather Index
GCEW	Global Commission on the Economics of Water
GDP	gross domestic product
GLDAS-2.2-DA	Global Land Data Assimilation System Version 2
GRACE	Gravity Recovery and Climate Experiment
GRACE-FO	Gravity Recovery and Climate Experiment Follow-On
HDI	Human Development Index
IPBES	Intergovernmental Science-Policy Platform on Biodiversity and Ecosystem Services
IPCC	Intergovernmental Panel on Climate Change
ISIMIP	Inter-Sectoral Impact Model Intercomparison Project
IWRM	integrated water resources management
JPL	Jet Propulsion Laboratory
LMDI	Logarithmic Mean Divisia Index
MAD	mean absolute deviation
MODIS	Moderate Resolution Imaging Spectroradiometer
NASA	National Aeronautics and Space Administration
NOAA	National Oceanic and Atmospheric Administration
NPP	net primary production

NRW	nonrevenue water
NTMs	nontariff measures
PPP	public-private partnership
SDGs	Sustainable Development Goals
SPEI	Standardized Precipitation Evapotranspiration Index
SRI	system of rice intensification
TV	total variance
TWS	terrestrial water storage
USDA	U.S. Department of Agriculture

All dollar amounts are in US dollars unless otherwise indicated.

CHAPTER 1

Trends in and Causes of Global Freshwater Decline

Key findings

- Global freshwater reserves, including the total amount of water stored above ground and in aquifers, have declined at an annual rate of 324 billion m^3—enough to meet the annual water needs of 280 million people—over the past two decades. The median basin-level reduction in freshwater is equivalent to about 3 percent of the annual renewable freshwater supply across all basins and 10 percent in arid basins already expericing drying.
- Although most of the world's dry areas continue to get drier and its wet areas continue to get wetter, dry areas are drying at a faster rate than wet areas are wetting, creating continental-scale mega-drying regions.
- Global warming, worsening droughts, and unsustainable water and land use all contribute to the reduction in global freshwater reserves.

Introduction

The year 2024 was the hottest year on Earth since records began in 1850 (NOAA 2025). Relentless heat waves scorched continents, and global temperatures surged to unprecedented levels, turning what was once extreme into the new normal. However, the consequences of rising temperatures extend beyond increased heat. As temperatures increase, evaporation from the Earth's surface accelerates, increasing the amount of water vapor in the atmosphere and disrupting the global water cycle. According to the Intergovernmental Panel on Climate Change, each additional increment of global warming intensifies water-related weather extremes, leading to more severe floods and prolonged droughts (Core Writing Team, Lee, and Romero 2023).

A reproducibility package is available for this book in the Reproducible Research Repository at https://reproducibility.worldbank.org/index.php/catalog/386.

A changing water cycle directly affects the availability of freshwater, as seen in shifts in freshwater reserves or terrestrial water storage (TWS). TWS is the total amount of water stored on land, including water found in glaciers, rivers, lakes, reservoirs, and aquifers and water held in soil moisture and vegetation. This chapter examines long-term TWS changes to assess the trends in freshwater availability, using more than two decades of data from the Gravity Recovery and Climate Experiment (GRACE) missions of the U.S. National Aeronautics and Space Administration (NASA) and the German Aerospace Center.

Numerous studies have identified TWS changes based on GRACE data (Reager et al. 2016; Rodell and Li 2023; Rodell et al. 2018, 2024; Scanlon et al. 2022). Burke et al. (2023) highlighted the long-term cumulative loss in TWS reported in various literature, and GCEW (2024) underscored the decline in TWS from 2002 to 2016.

Although regional trends are relatively well characterized in existing studies, uncertainties remain regarding whether these trends are robust, what the underlying drivers are, and whether the global continents, as a whole, are gaining or losing freshwater from or to oceans (Kim et al. 2019). Additionally, most studies are conducted at lower-resolution grid cells of 3° × 3°, meaning Earth's surface is divided into squares approximately 330 km by 330 km at the equator, akin to the size of a small country. Although this grid captures broad patterns, it limits detection of finer-scale changes, such as those occurring in smaller watersheds or local aquifers.

This chapter provides a comprehensive assessment of TWS changes over the past two decades, building and improving on the findings of previous studies. It presents the latest long-term trends in TWS from April 2002 to April 2024 on both global and regional scales. Additionally, the resolution of satellite observations has been enhanced from 3° × 3° to 0.25° × 0.25° (approximately 25 km by 25 km at the equator), allowing for more precise characterization of TWS changes at administrative boundaries, such as countries, states, and even counties. This finer resolution is integrated with socioeconomic data to uncover the drivers of TWS changes using novel methods.

Findings from the analysis reveal that the global continents (all land excluding Antarctica and Greenland) have undergone unprecedented rates of TWS loss since 2002—a phenomenon termed *continental drying* in this report. Although most of the world's dry areas continue to get drier and its wet areas continue to get wetter, dry areas are drying at a faster rate than wet areas are wetting. The rapid expansion of dry areas has led to the formation of mega-drying regions by connecting previously identified drying hot spots.

Using global hydrological models and econometric analyses, this chapter provides empirical evidence that human water and land use activities have significantly influenced changes in global freshwater reserves. Unregulated and poorly managed extraction of surface water and groundwater, combined with deforestation and wetland degradation, has intensified water scarcity. Although the crisis of our warming planet is fundamentally a water crisis, ineffective water management has exacerbated it.

Continental drying as seen from space

Monitoring changes in the availability of freshwater is critical for water resources management. However, tracking how water moves and is stored on Earth has been notoriously difficult, especially for groundwater, which represents 97 percent of Earth's unfrozen freshwater. In 2002, NASA and the German Aerospace Center launched a satellite mission, GRACE, which operated from 2002 to 2017, followed by its successor mission, the GRACE Follow-On (GRACE-FO), launched in 2018. By monitoring small but continuous changes in Earth's gravity field driven by the movement of water, GRACE and GRACE-FO (hereafter, GRACE) have allowed the measurement of how freshwater reserves vary in space and time across the entire planet (refer to box 1.1).

BOX 1.1

A scale in the sky: Measuring change in Earth's freshwater resources

The Gravity Recovery and Climate Experiment (GRACE, 2002–17) and its successor, the GRACE Follow-On (GRACE-FO, 2018–present) missions use two satellites that orbit the Earth and measure the distance between them with high precision, down to micrometers (a micrometer is 1/50th the width of a human hair). As the pair of satellites circle Earth, their positions are changed slightly by the amount of water mass in the region below. For example, when the front satellite approaches a large water reservoir, it accelerates because it is pulled slightly by the reservoir's gravity, changing the distance between the two satellites. When it has passed over the reservoir, it decelerates, and the velocity of the second satellite increases, again changing the distance between the two satellites. The next month, if the reservoir has lost water, the overall pull by the reservoir will be smaller; thus, the satellites' acceleration and deceleration will be smaller. The difference between the two months' measurements is then

(continued)

BOX 1.1

A scale in the sky: Measuring change in Earth's freshwater resources *(continued)*

converted into water mass units, after accounting for the other influences. By monitoring these small but continuous changes in Earth's gravity field, GRACE functions like a scale in the sky, enabling researchers to observe fluctuations in terrestrial water storage (TWS).

Although GRACE data are powerful, they have some limitations. First, although the satellite observations provide valuable large-scale measurements of TWS, their inherent resolution can be too coarse for practical regional water management applications. Thus, the regional analysis in this report uses higher-resolution (0.25° × 0.25°) estimates of TWS from the GRACE-assimilated National Aeronautics and Space Administration Global Land Data Assimilation System Version 2 model after additional bias correction (refer to the online technical appendixes).[a] Second, although GRACE offers the only direct observational data on total freshwater reserves, its relatively short observation period limits assessment of longer-term trends. However, recent studies incorporating extended data sets have identified declines in various components of TWS over longer time periods—including the long-term depletion of soil moisture (Seo et al. 2025), groundwater (Wada et al. 2010), surface water (Pekel et al. 2016), and lake storage (Yao et al. 2023)—reinforcing the findings of this study. Third, the GRACE missions observe monthly changes in TWS—not the absolute amount of TWS. Understanding the absolute amount of freshwater on and below the land surface would require an unprecedented level of exploration of Earth's shallow crustal water environment.

a. The technical appendixes are available online at https://hdl.handle.net/10986/43683.

Global land is losing freshwater to oceans

Globally, a key finding from the analyses of GRACE data is the consistent net loss of freshwater reserves at an annual rate of 324 billion m^3—enough to meet the annual water needs of 280 million people—over the past two decades (refer to figure 1.1). A long-term deficit in freshwater reserves indicates that water losses from land consistently exceed gains from precipitation, leading to a continuous depletion of freshwater reserves. If this trend continues, it will render the system unsustainable. The water lost from land ultimately flows into oceans, contributing to mass-driven sea level rise. For more details of the analysis, refer to Chandanpurkar et al. (2025).

Global freshwater availability has persistently declined over the past two decades.

FIGURE 1.1 Global freshwater loss to oceans, 2002–24

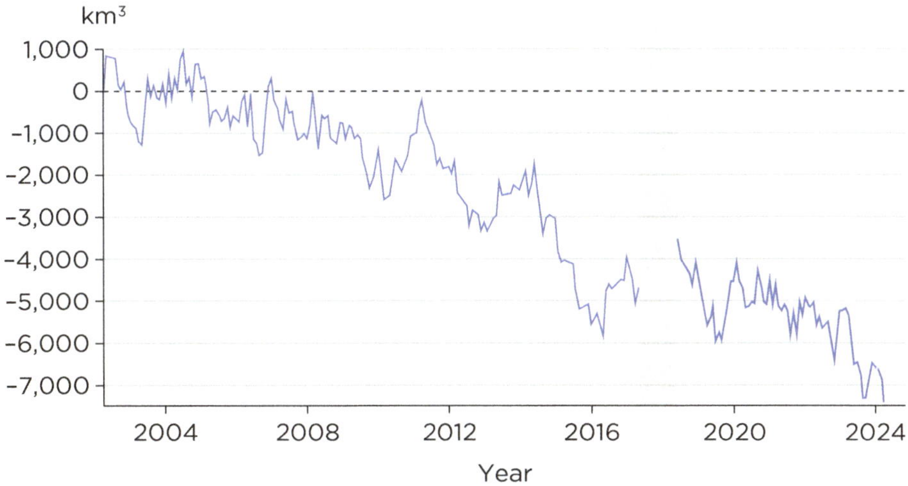

Source: Adapted from Chandanpurkar et al. 2025.
Note: This figure depicts the long-term loss of the global land water mass, measured by the net change in globally integrated, deseasoned TWS data from April 2002 to April 2024. TWS data are from the National Aeronautic and Space Administration's GRACE satellite mission, which operated from 2002 to 2017, and its successor mission, GRACE-FO, launched in 2018. The gray band indicates the gap period between the GRACE and GRACE-FO missions. GRACE = Gravity Recovery and Climate Experiment; GRACE-FO = Gravity Recovery and Climate Experiment Follow-On; km^3 = cubic kilometers; TWS = terrestrial water storage.

The emergence of continental-scale mega-drying regions

Beyond the global average, the trend varies across regions. Analyzing TWS trends at each land location reveals several hot spots of pronounced freshwater decline (refer to map 1.1). Some of these hot spots are well documented (for example, Rodell et al. 2018), including glaciers in Alaska, Canada, Central Asia, Patagonia, and the Himalayas, and the nonglaciated regions in northern China, northern India, the Middle East, and the southwestern United States. In recent years, however, the decline in TWS has also been seen in Central America, most of Europe, and high-latitude but nonglaciated parts of Eurasia and North America.

A major finding of this report is that the land between the previously identified hot spots is also consistently losing TWS, forming what is described in this report as continental-scale mega-drying regions, as illustrated in map 1.1a. These regions include (1) Alaska and western and northern Canada, (2) northern Russian Federation, (3) Central America and southwestern North America, and (4) the vast three-continent landmass that includes Europe, the Middle East and North Africa, Central Asia, South Asia (except peninsular India), Southeast Asia, and northern China.

The large-scale continental drying is also evident in the zonal plot in map 1.1b, which shows that all latitudes, except for the tropics between 10°S and 20°N, exhibit negative TWS trends, with or without accounting for the presence of glaciers. The trends have persisted for the past 22 years, showing little sensitivity to a lengthening GRACE record.

Rapid expansion of dry areas has led to the emergence of continental-scale mega-drying regions.

MAP 1.1 Emergence of continental-scale mega-drying regions

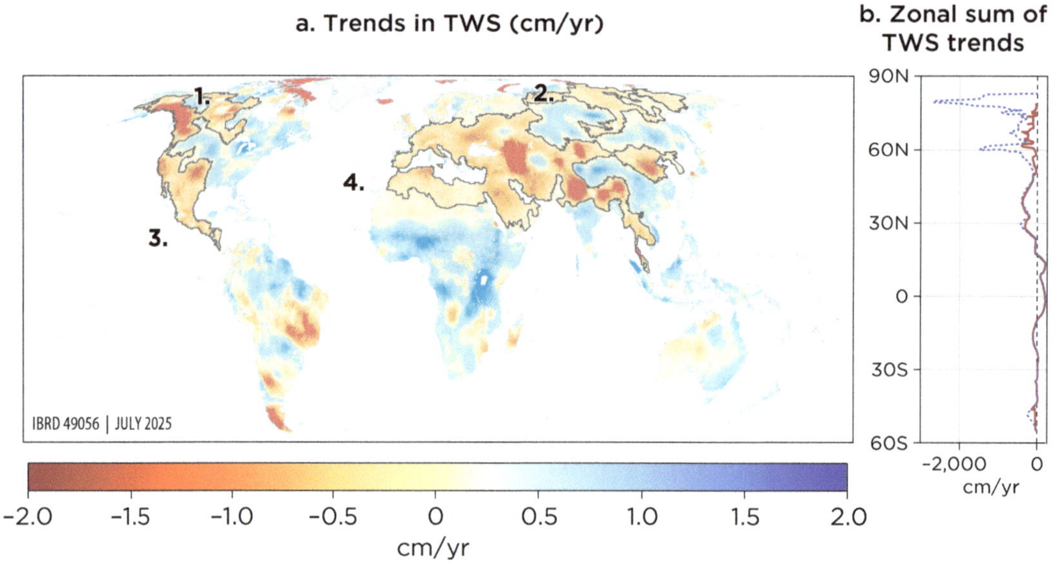

Source: Chandanpurkar et al. 2025.
Note: In panel a, red indicates regions that have experienced a persistent loss of freshwater reserves over the past two decades, and blue indicates areas that have consistently gained freshwater reserves during the same period. Intensity of color corresponds to the magnitude of water loss or gain. The numbers on the map indicate the mega-drying regions: 1 = Alaska and western and northern Canada; 2 = northern Russian Federation; 3 = Central America and southwestern North America; (4) Central Asia, northern China, Europe, the Middle East and North Africa, South Asia (except Peninsular India), and Southeast Asia. The zonal plot in panel b shows the sum of TWS across parallel lines, for all regions (blue dotted line) and for nonglaciated regions (red). It shows that, with the exception of the tropics between 10S and 20N, all latitudes exhibit a net decline in freshwater reserves, even when excluding glaciers and ice caps. cm/yr = centimeters per year; TWS = terrestrial water storage.

This persistent drying is accompanied by increased area under dry TWS anomalies and extremes. The locations experiencing below-average monthly TWS have been increasing by an average of 831,600±69,100 km^2 per year—more than the size of the Amazon River Basin.[1] The regions facing dry extremes—defined as below-average monthly TWS values that exceed 1 local standard deviation—has also grown at a similar rate, by an average of 845,000 ± 122,600 km^2 per year. This trend is primarily driven by drying in nonglaciated regions, which make up 72 percent of the area under dry anomalies and 81 percent of the area experiencing dry extremes. In contrast, both the areas becoming wetter and the regions experiencing wet extremes have decreased over the past 22 years (refer to box 1.2).

BOX 1.2

Drenched yet dry: Wetting and economic water scarcity

Although the continents are, on average, losing water, there are both wetting and drying hot spots—places that have experienced persistent increasing and decreasing amounts, respectively, of terrestrial water storage (TWS). Notably, wetting trends have been observed in eastern Australia, eastern central China, the northern Great Plains, northern North America, and the Okavango Delta in Southern Africa. These wetting trends were identified in Rodell et al. (2018) but have grown weaker. There are also wetting regions in which the magnitude of the latest trends is more pronounced than in the study period in Rodell et al. (2018). These regions include humid regions in central Africa, the Amazon, central India, Indonesia, and eastern North America.

Sub-Saharan Africa is the only region that shows an overwhelming increase in TWS (Rodell and Li 2023; Scanlon et al. 2022). This increase is largely due to a pronounced recent increase in precipitation in eastern and northern Sub-Saharan Africa that is likely due to natural climate variability patterns such as the El Niño–Southern Oscillation and the Indian Ocean Dipole, which caused widespread flooding and affected more than 2.8 million people (Wainwright et al. 2021). Because these natural climate variability patterns are tightly linked to the regional ocean temperature, their frequency and intensity are likely to increase with continued warming of oceans. This extreme wetting somewhat masked declining TWS because of groundwater and surface water consumption in the region.

(continued)

BOX 1.2

Drenched yet dry: Wetting and economic water scarcity *(continued)*

Although many parts of Sub-Saharan Africa do not face physical water scarcity, economic water scarcity remains a significant challenge because of inadequate infrastructure. Consequently, only 32 percent of the region's population has access to safely managed drinking water—that is, drinking water that is accessible on the premises, available when needed, and free from contamination. The situation is even more severe in rural areas, where access dropped to just 16 percent in 2022.

Inadequate water storage exacerbates seasonal water scarcity, a form of economic water scarcity. For instance, Cherrapunji, India, receives high annual rainfall, exceeding 11,000 mm, making it one of the wettest places on Earth. However, the vast majority of this rainfall occurs during the monsoon season (June to September). Because it has insufficient infrastructure for capturing and storing water, the city struggles to balance water supply and demand between wet and dry seasons, often facing severe shortages outside the monsoon months.

Relative importance of the drying trend

How significant are these TWS trends at the local level? To answer this question, the analysis for this report aggregates TWS trends across 4,639 river basins around the world and compares them with the annual renewable water supply in each basin, using enhanced-resolution GRACE data. The renewable water supply represents the remaining precipitation after accounting for evapotranspiration losses and the environmental flows necessary to sustain ecosystem health and the natural water cycle. Map 1.2 illustrates the magnitude of TWS trends as a share of the annual renewable water supply.

Globally, the median long-term TWS trend magnitude is about 3 percent of annual freshwater supply of all basins, 5 percent of the basins that are drying, and 2 percent of the basins that are wetting. The median TWS trend is 8 percent of annual renewable water supply for arid basins and 10 percent for basins that are arid and experiencing drying. The latter include basins in the southern midlatitudes in Australia, South Africa, and South America; in the Middle East and North Africa; and in the southwestern United States.

Continental drying exacerbates water scarcity in areas that can least afford it.

MAP 1.2 The median long-term TWS trend as a share of annual renewable freshwater supply, 2003–24

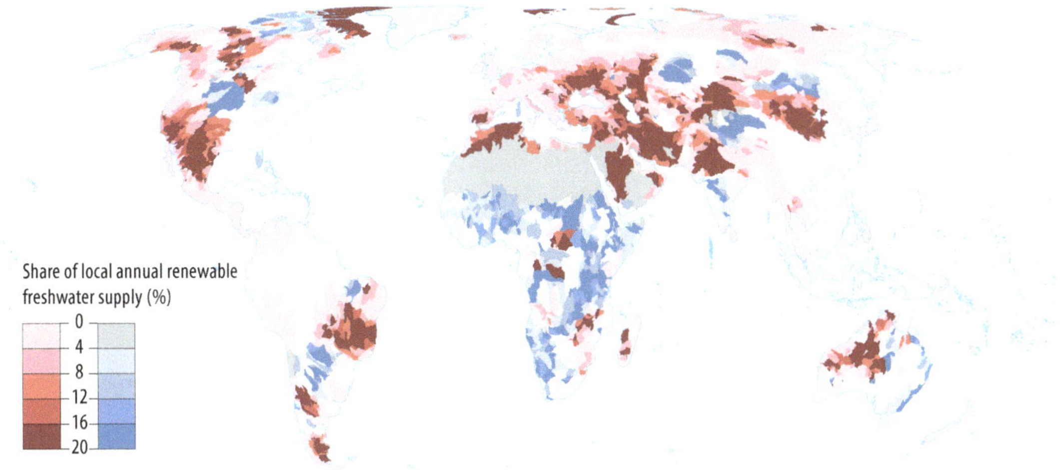

IBRD 49077 | AUGUST 2025

Source: Chandanpurkar et al. 2025.
Note: Red indicates drying regions, and blue indicates wetting regions. The analysis ignores hyperarid desert regions (aridity index < 0.03), where TWS is low. TWS = terrestrial water storage.

Notably, TWS values represent long-term trends, whereas water supply values are annual. A river basin's TWS trend would not be negative if its renewable water supply consistently exceeded water demand and losses. The basins highlighted in red on map 1.2 indicate regions where water losses and demands have persistently outpaced renewable supply, making it increasingly difficult to offset water storage deficits.

Drivers of continental drying

What causes the decline in freshwater reserves? Findings from previous studies and analysis for this report indicate that the global trend of decreasing TWS can predominantly be attributed to global warming, the increasing severity and duration of droughts, and human activities related to water consumption and land use change.

Snow and glacier melting substantially affect TWS in the context of global warming (Immerzeel et al. 2020). Moreover, studies have shown the significant impacts of droughts on TWS (Anyah et al. 2018; Castle et al. 2014;

Famiglietti et al. 2011; Liu et al. 2022). During a drought, rainfall deficits decrease runoff, soil moisture, and groundwater recharge, and rising temperatures increase atmospheric evaporative demand, leading to the loss of water from the soil. The severity of droughts has worsened in the past five years (Büntgen et al. 2021; NOAA 2025; Rodell et al. 2024; Rodell and Li 2023).

Additionally, earlier studies highlighted the complex interactions between human activity and climate change. For example, during periods of drought, when water resources become scarce, the interplay between climatic factors and human activities can accelerate the use of groundwater and surface water, leading to reductions in groundwater and lake water storage (Famiglietti et al. 2011; Wada et al. 2010; Yao et al. 2023).

Human activities, independent of reaction to droughts, also play a significant role in the depletion of freshwater resources through excessive water withdrawal and changes in land use. Groundwater, which provides on-demand local access to water, serves as a prime example of a common pool resource, which is characterized by its shared nature and vulnerability to overuse (Gordon 2000; Hardin 1968). Without proper water management, including formal regulations and well-defined ownership rights, groundwater often becomes overexploited, even during periods without drought (Brozović, Sunding, and Zilberman 2010; Edwards et al. 2016; Pfeiffer and Lin 2014). Overexploitation can deplete the resource faster than it is naturally replenished, ultimately leading to declining water tables and drying wells. Alarmingly, approximately half of the world's major aquifers are showing signs of rapid depletion (Richey et al. 2015; Rodell et al. 2018).

Deforestation presents another significant threat to global water resources. It can reduce precipitation, soil moisture, and the potential to recharge groundwater (Smith et al. 2023). GCEW (2024) highlights how deforestation and other land use changes can disrupt the water cycle and exacerbate local water scarcity.

Attributing TWS changes to specific drivers can be challenging, but examining the sources of TWS loss provides insights into the significance of various factors. This report estimates that annual freshwater loss in nonglaciated drying continental regions is increasing each year and has now exceeded the loss from melting glaciers and ice caps (excluding Antarctica and Greenland). Combining GRACE data with the global hydrological model WaterGAP 2.2d, analysis for this report estimates that in nonglaciated drying regions the single largest contributor to water storage loss comes from the depletion of groundwater (68 percent), followed by depletion of surface water (18 percent), soil moisture (9 percent), and snow water (5 percent). Detailed information on global hydrological modeling can be found in technical appendix A (online).[2]

The following sections delve deeper into the relationship between human activities and freshwater storage changes on a global scale, in particular by integrating down-scaled TWS data with socioeconomic indicators. Findings from these analyses highlight the link between anthropogenic pressures—such as agricultural practices, deforestation, and inadequate water resources management—and the decline in freshwater resources. These findings underscore the significant opportunity to implement sustainable water management strategies to slow the pace of continental drying.

Land use change

Human activities have profoundly transformed the global landscape, often reshaping natural ecosystems to meet agricultural, industrial, and urban demands. Irrigation has boosted yields on rainfed land and increased food supplies (Faurès, Hoogeveen, and Bruinsma 2002; Sarsons 2015), but it is also the largest consumer of water, accounting for about 70 percent of the freshwater diverted by human activity, with almost 50 percent supplied from groundwater (Siebert et al. 2010). Numerous studies have documented the correlation between historical irrigation practices and substantial groundwater depletion, especially in regions heavily reliant on agriculture (Dalin et al. 2017; GCEW 2024).

Forests function as natural reservoirs and filters, playing a crucial role in storing and purifying water. However, global deforestation has disrupted these critical hydrological processes (GCEW 2024). From 2010 to 2015, tropical forests shrank by 5.5 million hectares annually while temperate forests expanded by 2.2 million hectares annually (Zhang and Wei 2021). Although the effects of deforestation on water yields—specifically, the volume of water in streamflow—are still a topic of debate (Filoso et al. 2017), studies have associated deforestation with approximately 4 percent of the recent drying observed in the Amazon (Staal et al. 2020) and with decreased access to clean drinking water, equivalent to a 9 percent decrease in rainfall, in Malawi (Mapulanga and Naito 2019).

Wetlands also play an essential role in natural water retention and purification (Bring et al. 2020; Lv et al. 2019; Pal and Talukdar 2018), yet the scale of wetland degradation is even more alarming. Since 2002, the global area of permanent wetlands has diminished by about 468,000 km^2, a 24 percent reduction, accounting for approximately 21 percent of all wetlands that have existed since 1700 (Fluet-Chouinard et al. 2023). The primary causes of these losses include wetland drainage for upland croplands (about 61.7 percent of total loss), conversion to flooded rice fields (18.2 percent), urban development (8.0 percent), forestry (4.7 percent), wetland cultivation (4.3 percent), pasture (2.0 percent), peat extraction (0.9 percent, according to Fluet-Chouinard et al. 2023), and the broad-scale disappearance of groundwater (Rohde et al. 2024).

Urban expansion frequently results in the conversion of natural and agricultural lands into urban spaces, diminishing forest coverage. In addition, urban development creates roads and buildings that prevent water from soaking into the ground, reducing natural groundwater recharge and soil moisture. Urban areas also experience increased groundwater abstraction for industrial and commercial activities, further depleting groundwater resources (Rodella, Zaveri, and Bertone 2023). High-resolution mapping of global land use reveals that urban areas have expanded by 50,000 km^2, a 6 percent increase since 2002. Although urban land accounts for only a small fraction of total land area, urbanization is anticipated to continue its upward trend, with 68 percent of the global population projected to reside in urban areas by 2050, up from 55 percent in 2018.

By overlaying land use and TWS data, this report's analysis indicates that the type of land use in 2002 significantly influenced the trajectory of local freshwater availability from 2003 to 2022 (refer to figure 1.2 and box 1.3).[3] Specifically, compared with barren land, a cell with an additional 1 percent of land, or approximately 25 km^2, allocated for urban use will experience an accelerated TWS depletion rate of 0.74 million tons per year over the next two decades. Furthermore, if a cell has an additional 1 percent of land removed from forests, it will experience an accelerated TWS depletion rate of 0.14 million tons annually, assuming all other factors remain constant. Similarly, a 1 percent reduction in wetland area is correlated with an accelerated TWS depletion rate of 0.19 million tons per year under the same conditions. Additionally, allocating an extra 1 percent of land to irrigation results in an accelerated TWS depletion rate of 0.48 million tons per year within that cell. These findings highlight the critical impact of land use changes on TWS depletion.

Land use decisions are a key driver of continental drying in nonglaciated areas.

FIGURE 1.2 Impact of 1 percent change in land use type in 2002 on TWS trends, 2003–24

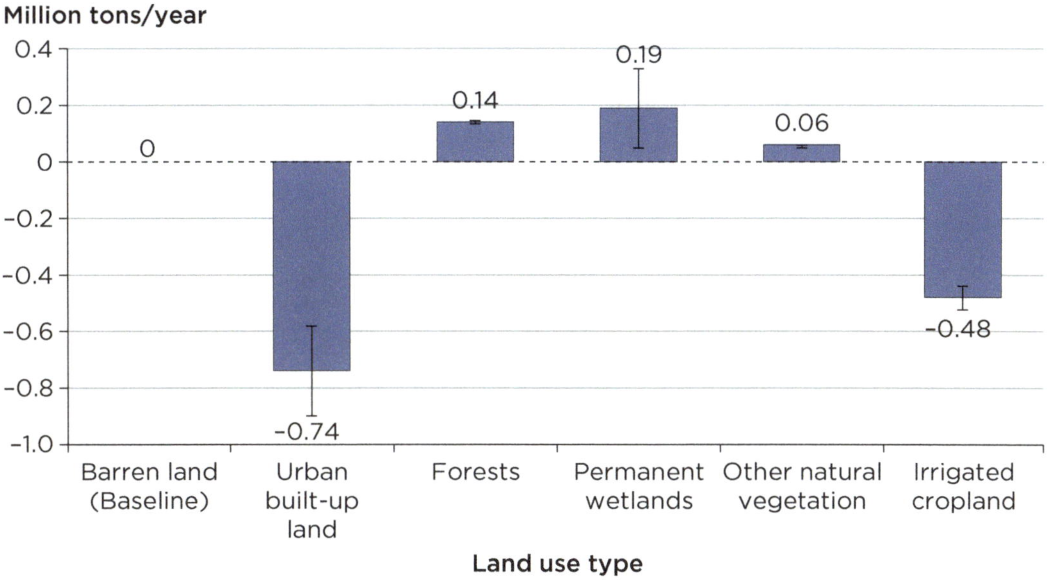

Source: World Bank.

Note: This figure shows the estimated impact of a 1 percent change in the respective land use type in 2002 (equivalent to approximately 25 km^2 on average) on the grid cell's TWS trend from 2003 to 2024, with barren land serving as the baseline. "Other natural vegetation" includes grasslands, savannas, and shrublands. TWS = terrestrial water storage.

BOX 1.3

Untangling the link: The impact of land use change on freshwater availability

Understanding the effects of land use on freshwater availability is not straightforward. Land use is an endogenous decision of society. Taking irrigation as an example, on the one hand, intensive irrigation and excessive water extraction can contribute to the depletion of terrestrial water storage (TWS). On the other hand, changes in water availability can influence the decisions for irrigation development and land use. For example, during prolonged droughts, local populations may respond to decreased water resources by converting more land to agriculture in an effort to offset the loss of productivity on existing

(continued)

BOX 1.3

Untangling the link: The impact of land use change on freshwater availability *(continued)*

cropland (Damania et al. 2017). Thus, using contemporaneous data on land use and TWS can capture both how initial land use decisions affect water resources and how changes in water availability influence land use.

To estimate the impact of land use on water resources, this report estimates the effects of initial land configurations in 2002 on subsequent TWS trends from 2003 to 2024. By focusing on land use data from this earlier time, land use decisions are plausibly exogenous—independent of and unaffected by the TWS outcomes in the decades that follow. This approach helps mitigate endogeneity issues, whereby variables may be influenced by the outcomes they aim to explain, and it thus provides a more accurate picture of how land use configurations influence TWS over time.

Pricing distortions

Accelerated TWS depletion in intensively irrigated cropland is significantly influenced by distortions in water and energy pricing. Water is rarely priced or metered for agricultural water users because irrigation has historically been subsidized to meet other policy objectives, including poverty alleviation, productivity gains, and food security. Evidence suggests that farmers perceive metering as a precursor to regulation or pricing, which increases input costs and may threaten competitiveness in regional and global value chains (Hellegers and Davidson 2024). Where water pricing exists, it often relies on fixed cost structures (Chakravorty, Dar, and Emerick 2023). Although fixed pricing offers a practical solution in areas lacking metering infrastructure, it does not encourage water conservation. As noted by Chakravorty, Dar, and Emerick (2023), only 26 of 80 countries with available water pricing data use volumetric pricing. When irrigation costs bear no relation to actual consumption, farmers are left with little motivation to adopt water-saving technologies or modify their farming practices in ways that could reduce water use.

In regions lacking effective agricultural water pricing, the cost of extracting groundwater aligns solely with the energy needed for pumping

(Badiani-Magnusson and Jessoe 2018; Burlig, Preonas, and Woerman 2020; Fishman et al. 2016). The absence of a price signal for water encourages consumers to extract until the marginal benefit of an additional unit of groundwater equals the net energy extraction costs. Consequently, energy subsidies—especially those for the electricity used in groundwater pumping—further diminish the cost of groundwater, exacerbating the problem of groundwater depletion (Aeschbach-Hertig and Gleeson 2012). Underpriced irrigation not only triggers overpumping but also incentivizes farmers to cultivate more water-intensive crops (Sayre and Taraz 2019; Sekhri 2011).

Research indicates that farmers are responsive to the cost of irrigation water. Badiani and Jessoe (2013) observed that a 10 percent increase in electricity subsidies resulted in a 6.6 percent rise in groundwater extraction in India. Similarly, Chakravorty, Dar, and Emerick (2023) demonstrated that water pricing is pivotal in the adoption of water conservation technologies. In Bangladesh, use of the water-saving technique known as alternate wetting and drying increased by 21 percent when villages transitioned from fixed charges to volumetric pricing. Even with the implementation of efficient irrigation technologies, pricing remains a critical element in water conservation. Fishman, Giné, and Jacoby (2023) found that, when a group of smallholders received a 90 percent subsidy on their drip irrigation installation, no reduction in groundwater pumping was observed after three years; given the minimal costs associated with pumping, farmers opted to maximize use and sell the excess water to neighboring landowners rather than conserve it.

This report uses annual energy inflation data from the World Bank's Global Consumer Price Index data set, covering 142 countries, and identifies a statistically significant negative correlation between energy prices and the rate of TWS depletion in countries with intensive irrigation, defined as those with more than 20 percent of cropland irrigated.[4] As shown in figure 1.3, in countries that depend heavily on irrigation, lower energy prices are associated with a faster decrease in water levels and higher energy prices with a slower depletion rate. In contrast, the impact of energy prices on TWS is statistically insignificant in countries with minimal irrigation. Technical appendix A (online) provides additional details regarding the data and methodology used in the analysis.

Notably, not all irrigation systems depend on groundwater pumping. Gravity-fed irrigation systems require no energy inputs for water transport. For these systems, fluctuations in energy prices may have little to no effect on TWS levels, which can lead to some noise in the data regarding the relationship between energy prices and water storage. Nevertheless, a negative correlation between energy prices and TWS levels persists, indicating that energy subsidies and the general underpricing of water incentivize groundwater extraction, ultimately contributing to the depletion of freshwater resources.

Addressing pricing distortions can help preserve freshwater resources.

FIGURE 1.3 **Impact of energy pricing on freshwater reserves in irrigation-intensive countries**

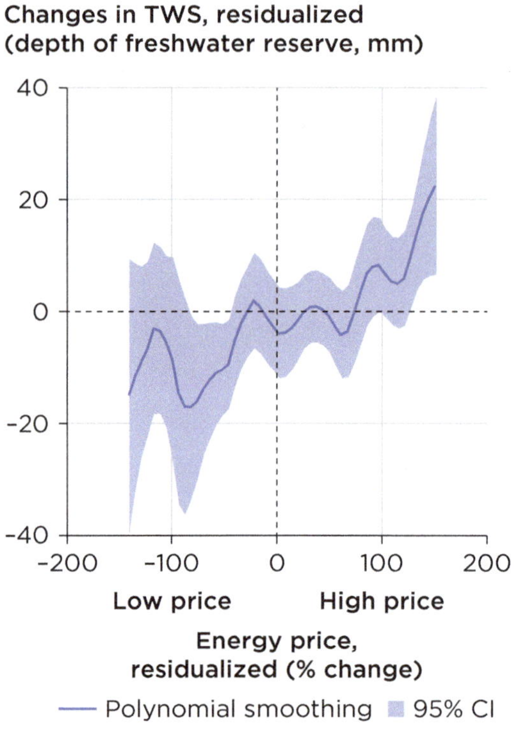

Source: World Bank.
Note: This figure illustrates the relationship between changes in energy prices and changes in freshwater reserves (measured by TWS). Both energy price and TWS level are residualized, controlling for annual average temperature and precipitation (and their polynomial functions up to the third order), country fixed effects, and year fixed effects. Irrigation-intensive countries are defined as countries having more than 20 percent of their cropland irrigated. The shading indicates 95 percent CIs, representing the precision of the estimates. CI = confidence interval; TWS = terrestrial water storage.

Lack of integrated water management

The previous section underscores that water outcomes are often shaped by decisions made outside of the water sector, particularly decisions regarding land use. Therefore, it is important to coordinate the management of land, water, and related resources while balancing the competing demands of various sectors, including agriculture, urban development, and conservation, to safeguard water resources.

The concept of managing water resources using integrated approaches—integrated water resources management (IWRM)—has been incorporated into the Sustainable Development Goals (SDGs) as reflected in SDG indicator 6.5.1,[5] which tracks the degree of IWRM implementation across countries. This indicator has been measured over three data collection rounds, covering 173 countries in 2017, 186 countries in 2020, and 183 countries in 2023. A country's IWRM score is determined by performance on four key dimensions: enabling environment (laws, policies, and plans supporting IWRM), institutions and participation (institutional capacity and stakeholder engagement), management instruments (data collection and planning tools), and financing (investments and funding for IWRM efforts and infrastructure). The scoring uses a scale ranging from 0 to 100, with higher scores reflecting more comprehensive and effective IWRM implementation. The data on the degree of IWRM implementation are collected from country self-assessments and validated through stakeholder consultations.

This report evaluates the relationship between IWRM scores and average TWS depletion rates over the three years after the survey. The findings reveal that inadequate IWRM implementation accelerates TWS depletion, whereas adequate IWRM implementation leads to more sustainable TWS outcomes, all else equal (refer to figure 1.4).[6] Specifically, a decline of 1 standard deviation in a country's IWRM score is, on average, associated with an accelerated TWS decline of 0.011 km^3 per year per cell, that is, 11 million tons of water per year per cell, or a 0.36 standard deviation change in the rate of depletion.

To avoid potential biases associated with self-reporting that could confound the estimation results, additional analysis was conducted to compare TWS trends of extremely poor performers with those of other performers.[7] The results show that countries with IWRM scores lower than 30 experienced an average TWS decline that occurred two to three times faster than that of the countries with more robust IWRM implementation. Overall, the evidence indicates that enhancing IWRM can play a pivotal role in mitigating water depletion and fostering environmental sustainability.

Strengthening integrated water resources management can help preserve freshwater resources

FIGURE 1.4 **Impact of IWRM on freshwater reserves**

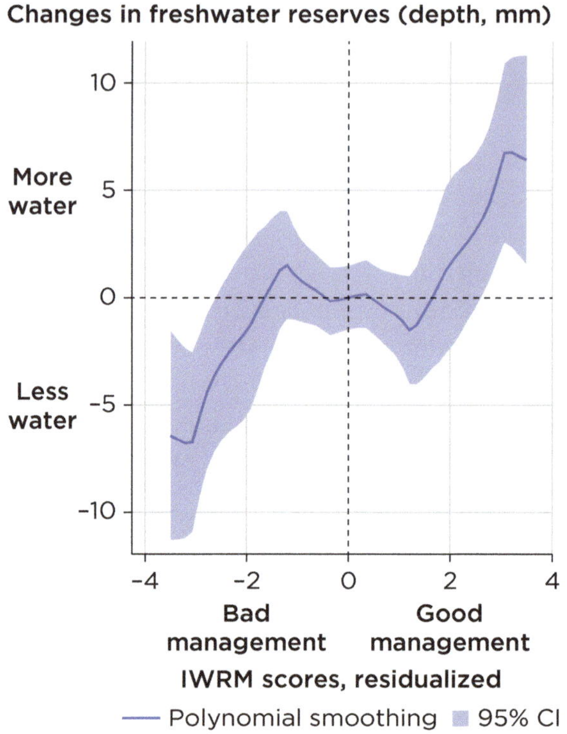

Source: World Bank.

Note: This figure illustrates the relationship between the score of IWRM and changes in freshwater reserves (measured by TWS). The x axis represents a country's degree of IWRM implementation for a given survey round; the y axis shows the corresponding TWS trend in the three-year period after the survey round. Both variables are residualized, controlling for the country's annual average temperature and precipitation (and their polynomial functions up to the third order), country fixed effects, and survey-year fixed effects. The unit of the y axis refers to the equivalent depth of TWS in millimeters (that is, the depth of water when it is uniformly distributed across each 0.5° grid cell). The shading indicates 95 percent CIs, representing the precision of the estimates. CI = confidence interval; IWRM = integrated water resources management; mm = millimeters; TWS = terrestrial water storage.

Conclusion

Using observational satellite data from GRACE, this chapter highlights an alarming trend of continental drying—a persistent decline in freshwater availability across vast landmasses. This drying is driven by a combination of

a changing climate and unsustainable water and land practices.
The potential risks of continental drying are profound: it can reduce agricultural productivity, heighten competition for water resources, and, in extreme cases, trigger ecosystem collapse and large-scale emigration. The next chapter explores the cascading impacts of continental drying on people, the economy, and the environment.

Notes

1. Areas under drying conditions (or dry anomalies) are spread across different regions each year.
2. Technical appendixes A through E are available online at https://hdl.handle.net/10986/43683.
3. The analysis controls for local precipitation and temperature influences as well as other unobservable and time-invariant location-specific characteristics. Regions where ice sheet and glacier melting have caused significant TWS decreases, such as Greenland, the Gulf of Alaska, and Patagonia, are removed from the sample. Refer to box 1.3 and technical appendix A (online) for details on the methodology.
4. Energy prices are lagged to address potential endogeneity between energy costs and the irrigation-driven demand for energy. The findings remain robust after controlling for variables such as temperature, precipitation, time-invariant country characteristics, and year fixed effects, along with alternative definitions of irrigation intensity. Specifically, 40 countries are classified as irrigation-heavy on the basis of the criterion that more than 20 percent of their cropland is primarily irrigated. Alternatively, a country is deemed irrigation-heavy if it has more irrigated cropland than rainfed cropland. The results are robust under both definitions.
5. UN SDG Indicator 6.5.1. 'Degree of Integrated Water Resources Management Implementation (0-100)," https://www.unwater.org/our-work/sdg-6-integrated-monitoring-initiative/indicator-651-degree-integrated-water-resources.
6. The analysis uses country-level panel data and controls for temperature, precipitation, and country and survey-year fixed effects.
7. In 2017, about 19 percent of countries were classified as having a low or very low level of IWRM implementation, with scores below 30. This percentage decreased to about 12 percent in 2020, and, by 2023, about 8 percent of surveyed countries remained in the low implementation category.

References

Aeschbach-Hertig, W., and T. Gleeson. 2012. "Regional Strategies for the Accelerating Global Problem of Groundwater Depletion." *Nature Geoscience* 5 (12): 853–61. https://doi.org/10.1038/ngeo1617.

Anyah, R. O., E. Forootan, J. L. Awange, and M. Khaki. 2018. "Understanding Linkages between Global Climate Indices and Terrestrial Water Storage Changes over Africa Using GRACE Products." *Science of The Total Environment* 635: 1405–16. https://doi.org/10.1016/j.scitotenv.2018.04.159.

Badiani-Magnusson, R., and K. Jessoe. 2018. "Electricity Prices, Groundwater, and Agriculture: The Environmental and Agricultural Impacts of Electricity Subsidies in India." In *Agricultural Productivity and Producer Behavior*, edited by W. Schlenker, 157–83. Chicago: University of Chicago Press.

Bring, A., L. Rosen, J. Thorslund, K. Tonderski, C. Åberg, I. Envall, and H. Laudon. 2020. "Groundwater Storage Effects From Restoring, Constructing or Draining Wetlands in Temperate and Boreal Climates: A Systematic Review Protocol." *Environmental Evidence* 9 (1): 26. https://doi.org/10.1186/s13750-020-00209-5.

Brozović, N., D. L. Sunding, and D. Zilberman. 2010. "On the Spatial Nature of the Groundwater Pumping Externality." *Resource and Energy Economics* 32 (2): 154–64.

Büntgen, U., O. Urban, P. J. Krusic, M. Rybníček, T. Kolář, T. Kyncl, A. Ač, et al. 2021. "Recent European Drought Extremes beyond Common Era Background Variability." *Rode* 14 (4): 190–96. https://doi.org/10.1038/s41561-021-00698-0.

Burke, M., J. M. Tront, K. N. Lyon, W. Rex, M. I. Castera Errea, M. C. Varughese, J. T. Newton, et al. 2023. *What the Future Has in Store: A New Paradigm for Water Storage—Overview for Policy Makers*. Washington, DC: World Bank. http://documents.worldbank.org/curated/en/099203002012336127.

Burlig, F., L. Preonas, and M. Woerman. 2020. "Groundwater, Energy, and Crop Choice." Working Paper, University of Chicago, Chicago.

Castle, S. L., B. F. Thomas, J. T. Reager, M. Rodell, S. C. Swenson, and J. S. Famiglietti. 2014. "Groundwater Depletion During Drought Threatens Future Water Security of the Colorado River Basin." *Geophysical Research Letters* 41 (16): 5904–11. https://doi.org/10.1002/2014GL061055.

Chakravorty, U., M. H. Dar, and K. Emerick. 2023. "Inefficient Water Pricing and Incentives for Conservation." *American Economic Journal: Applied Economics* 15 (1): 319–50.

Chandanpurkar, Hrishikesh A., James S. Famiglietti, Kaushik Gopalan, David N. Wiese, Yoshihide Wada, Kaoru Kakinuma, John T. Reager, and Fan Zhang. 2025. "Unprecedented Continental Drying, Shrinking Freshwater Availability, and Increasing Land Contributions to Sea Level Rise." Science Advances, 11: eadx0298. https://doi.org/10.1126/sciadv.adx0298. Background paper for this report.

Core Writing Team, Hoesung Lee, and José Romero, eds. 2023. *Climate Change 2023: Synthesis Report*. Geneva: Intergovernmental Panel on Climate Change. https://doi.org/10.59327/IPCC/AR6-9789291691647.

Dalin, C., Y. Wada, T. Kastner, and M. J. Puma. 2017. "Groundwater Depletion Embedded in International Food Trade." *Nature* 543, 700–4. https://doi.org/10.1038/nature21403.

Damania, Richard, Sébastien Desbureaux, Aude-Sophie Rodella, Jason Russ, and Esha Zaveri. 2017. *Uncharted Waters: The New Economics of Water Scarcity and Variability*. Washington, DC: World Bank.

Edwards, E. C., O. Cristi, G. Edwards, and G. D. Libecap. 2016. "An Illiquid Market in the Desert: Estimating the Cost of Water Trade Restrictions in Northern Chile." *Environment and Development Economics* 23: 615–34.

Famiglietti, J. S., M. Lo, S. L. Ho, J. Bethune, K. J. Anderson, T. H. Syed, S. C. Swenson, C. R. de Linage, and M. Rodell. 2011. "Satellites Measure Recent Rates of Groundwater Depletion in California's Central Valley." *Geophysical Research Letters* 38 (3): L03403. https://doi.org/10.1029/2010GL046442.

Faurès, J. M., J. Hoogeveen, and J. Bruinsma. 2002. *The FAO Irrigated Area Forecast for 2030*. Rome: Food and Agriculture Organization of the United Nations.

Filoso, S., M. O. Bezerra, K. C. Weiss, and M. A. Palmer. 2017. "Impacts of Forest Restoration on Water Yield: A Systematic Review." *PLOS One* 12 (8): e0183210.

Fishman, R., X. Giné, and H. G. Jacoby. 2023. "Efficient Irrigation and Water Conservation: Evidence from South India." *Journal of Development Economics* 162: 103051. https://doi.org/10.1016/j.jdeveco.2023.103051.

Fishman, R., U. Lall, V. Modi, and N. Parekh. 2016. "Can Electricity Pricing Save India's Groundwater? Field Evidence from a Novel Policy Mechanism in Gujarat." *Journal of the Association of Environmental and Resource Economists* 3 (4): 819–55.

Fluet-Chouinard, E., B. D. Stocker, Z. Zhang, A. Malhotra, J. R. Melton, B. Poulter, J. O. Kaplan, et al. 2023. "Extensive Global Wetland Loss over the Past Three Centuries." *Nature* 614 (7947): 281–6. https://doi.org/10.1038/s41586-022-05572-6.

GCEW (Global Commission on the Economics of Water). 2024. *The Economics of Water: Valuing the Hydrological Cycle as a Global Public Good*. Colombo, Sri Lanka: GCEW, International Water Management Institute.

Gordon, H. S. 2000. "The Economic Theory of a Common-Property Resource: The Fishery." In *Classic Papers in Natural Resource Economics*, edited by Chennat Gopalakrishnan, 178–203. London: Palgrave Macmillan.

Hardin, G. 1968. "The Tragedy of the Commons." *Science* 162: 1243–8.

Hellegers, Petra, and Brian Davidson. 2024. "Why Irrigation Water Pricing Is Difficult." *Water Economics and Policy* 10 (1): 2371001. https://doi.org/10.1142/S2382624X23710017.

Immerzeel, W. W., A. F. Lutz, M. Andrade, A. Bahl, H. Biemans, T. Bolch, S. Hyde, et al. 2020. "Importance and Vulnerability of the World's Water Towers." *Nature* 577: 364–69. https://doi.org/10.1038/s41586-019-1822-y.

Kim, J.-S., K.-W. Seo, T. Jeon, J. Chen, and C. R. Wilson. 2019. "Missing Hydrological Contribution to Sea Level Rise." *Geophysical Research Letters* 46 (21): 12049–55. https://doi.org/10.1029/2019gl085470.

Liu, P.-W., J. S. Famiglietti, A. J. Purdy, K. H. Adams, A. L. McEvoy, J. T. Reager, R. Bindlish, D. N. Wiese, C. H. David, and M. Rodell. 2022. "Groundwater Depletion in California's Central Valley Accelerates during Megadrought." *Nature Communications* 13 (1): 7825. https://doi.org/10.1038/s41467-022-35582-x.

Lv, J., W. Jiang, W. Wang, Z. Wu, Y. Liu, X. Wang, and Z. Li. 2019. "Wetland Loss Identification and Evaluation Based on Landscape and Remote Sensing Indices in Xiong'an New Area." *Remote Sensing* 11: 2834. https://doi.org/10.3390/rs11232834.

Mapulanga, A. M., and H. Naito. 2019. "Effect of Deforestation on Access to Clean Drinking Water." *Proceedings of the National Academy of Sciences* 116 (17): 8249–54. https://doi.org/10.1073/pnas.1814970116.

NOAA (National Aeronautic and Atmospheric Administration). 2025. "2024 Was the World's Warmest Year on Record." News release, January 10, 2025. https://www.noaa.gov/news/2024-was-worlds-warmest-year-on-record.

Pal, S., and S. Talukdar. 2018. "Drivers of Vulnerability to Wetlands in Punarbhaba River Basin of India-Bangladesh." *Ecological Indicators* 93: 612–26. https://doi.org/10.1016/J.ECOLIND.2018.05.043.

Pekel, J.-F., A. Cottam, N. Gorelick, and A. S. Belward. 2016. "High-Resolution Mapping of Global Surface Water and Its Long-Term Changes." *Nature* 540 (7633): 418–22. https://doi.org/10.1038/nature20584.

Pfeiffer, L., and C.-Y. C. Lin. 2014. "Does Efficient Irrigation Technology Lead to Reduced Groundwater Extraction? Empirical Evidence." *Journal of Environmental Economics and Management* 67 (2): 189–208. https://doi.org/10.1016/j.jeem.2013.12.002.

Reager, J. T., A. S. Gardner, J. S. Famiglietti, D. N. Wiese, A. Eicker, and M.-H. Lo. 2016. "A Decade of Sea Level Rise Slowed by Climate-Driven Hydrology." *Science* 351 (6274): 699–703. https://doi.org/10.1126/science.aad8386.

Richey, A. S., B. F. Thomas, M. H. Lo, J. T. Reager, J. S. Famiglietti, K. Voss, S. Swenson, and M. Rodell. 2015. "Quantifying Renewable Groundwater Stress With GRACE." *Water Resources Research* 51: 5217–38.

Rodell, M., A. Barnoud, F. R. Robertson, R. P. Allan, A. Bellas-Manley, M. G. Bosilovich, D. Chambers, et al. 2024. "An Abrupt Decline in Global Terrestrial Water Storage and Its Relationship with Sea Level Change." *Surveys in Geophysics* 45 (6): 1875–1902. https://doi.org/10.1007/s10712-024-09860-w.

Rodell, M., J. S. Famiglietti, D. N. Wiese, J. T. Reager, H. K. Beaudoing, F. W. Landerer, and M.-H. Lo. 2018. "Emerging Trends in Global Freshwater Availability." *Nature* 557 (7707): 651–9. https://doi.org/10.1038/s41586-018-0123-1.

Rodell, M., and B. Li. 2023. "Changing Intensity of Hydroclimatic Extreme Events Revealed by GRACE and GRACE-FO." *Nature Water* 1: 241–8. https://doi.org/10.1038/s44221-023-00040-5.

Rodella, A., E. Zaveri, and F. Bertone. 2023. *The Hidden Wealth of Nations: The Economics of Groundwater in Times of Climate Change*. Washington, DC: World Bank.

Rohde, M. M., C. M. Albano, X. Huggins, K. R. Klausmeyer, C. Morton, A. Sharman, E. Zaveri, et al. 2024. "Groundwater-Dependent Ecosystem Map Exposes Global Dryland Protection Needs." *Nature* 632 (8023): 101–7. https://doi.org/10.1038/s41586-024-07702-8.

Sarsons, H. 2015. "Rainfall and Conflict: A Cautionary Tale." *Journal of Development Economics* 115: 62–72.

Sayre, S. S., and V. Taraz. 2019. "Groundwater Depletion in India: Social Losses From Costly Well Deepening." *Journal of Environmental Economics and Management* 93: 85–100.

Scanlon, B. R., A. Rateb, A. Anyamba, S. Kebede, A. M. MacDonald, M. Shamsudduha, J. Small, A. Sun, R. G. Taylor, and H. Xie. 2022. "Linkages between GRACE Water Storage, Hydrologic Extremes, and Climate Teleconnections in Major African Aquifers." *Environmental Research Letters* 17 (1): 014046. https://doi.org/10.1088/1748-9326/ac3bfc.

Sekhri, S. 2011. "Public Provision and Protection of Natural Resources: Groundwater Irrigation in Rural India." *American Economic Journal: Applied Economics* 3 (4): 29–55.

Seo, K.-W., D. Ryu, T. Jeon, K. Youm, J.-S. Kim, E. H. Oh, J. Chen, J. S. Famiglietti, and C. R. Wilson. 2025. "Abrupt Sea Level Rise and Earth's Gradual Pole Shift Reveal Permanent Hydrological Regime Changes in the 21st Century." *Science* 387 (6741): 1408–13. https://doi.org/10.1126/science.adq6529.

Siebert, S., J. Burke, J. M. Faures, K. Frenken, J. Hoogeveen, P. Döll, and F. T. Portmann. 2010. "Groundwater Use for Irrigation—A Global Inventory." *Hydrology and Earth System Sciences* 14 (10): 1863–80.

Smith, C., E. Robertson, R. Chadwick, D. I. Kelley, A. P. K. Argles, C. A. S. Coelho, D. C. De Souza, et al. 2023. "Observed and Simulated Local Climate Responses to Tropical Deforestation." *Environmental Research Letters* 18 (10): 104004. https://doi.org/10.1088/1748-9326/acf0da.

Staal, A., B. M. Flores, A. P. D. Aguiar, J. H. Bosmans, I. Fetzer, and O. A. Tuinenburg. 2020. "Feedback between Drought and Deforestation in the Amazon." *Environmental Research Letters* 15 (4): 044024.

Wada, Y., L. P. H. Van Beek, C. M. Van Kempen, J. W. T. M. Reckman, S. Vasak, and M. F. P. Bierkens. 2010. "Global Depletion of Groundwater Resources." *Geophysical Research Letters* 37 (20): 2010GL044571. https://doi.org/10.1029/2010GL044571.

Wainwright, C. M., D. L. Finney, M. Kilavi, E. Black, and J. H. Marsham. 2021. "Extreme Rainfall in East Africa, October 2019–January 2020 and Context under Future Climate Change." *Weather* 76 (1): 26–31. https://doi.org/10.1002/wea.3824.

Yao, F., B. Livneh, B. Rajagopalan, J. Wang, J.-F. Crétaux, Y. Wada, and M. Berge-Nguyen. 2023. "Satellites Reveal Widespread Decline in Global Lake Water Storage." *Science* 380 (6646): 743–9. https://doi.org/10.1126/science.abo2812.

Zhang, M., and X. Wei. 2021. "Deforestation, Forestation, and Water Supply." *Science 371*: 990–1. https://doi.org/10.1126/science.abe7821.

CHAPTER 2

Cascading Effects: The Impact of Continental Drying on People, Food, and the Planet

Key findings

- Water scarcity significantly affects jobs and income. Between 2005 and 2018, droughts in Sub-Saharan Africa left 600,000–900,000 people jobless each year, a range equivalent to 7–10 percent of annual job creation in the region.
- Drying worsens the negative impact of warming on agricultural productivity. In the long term, the combined effects of drying and warming could push societies toward a tipping point where damage accelerates rapidly and adaptation becomes increasingly difficult.
- Beyond connections throughout the physical hydrological system, increased agricultural trade amplifies water dependencies among countries, transforming local water scarcity and water mismanagement into a collective global risk.
- The reduction in terrestrial water storage significantly raises the probability and intensity of wildfires worldwide. Global biodiversity hot spots are particularly sensitive to drying-induced wildfires.

Introduction

Water is essential for life and is intricately linked to all aspects of economic well-being (Zhang and Borja-Vega 2024). Historically, favorable rainfall has been a key driver of agricultural success, fueling the growth of early civilizations. Reliable water sources enabled irrigation, transportation, and trade, fostering complex societies and thriving cities. Conversely, devastating droughts, such as those that struck ancient Mesopotamia and the Maya, severely weakened agrarian economies and contributed to the collapse of empires (Boccaletti 2021).

A reproducibility package is available for this book in the Reproducible Research Repository at https://reproducibility.worldbank.org/index.php/catalog/386.

Given the critical role of water, the continental-scale reduction in freshwater reserves poses a significant challenge to future societies and ecosystems. An increasing body of research highlights the multifaceted economic and social consequences of water scarcity. Rainfall shocks, particularly droughts, reduce agricultural productivity and slow economic growth (Brown et al. 2013; Damania, Desbureaux, and Zaveri 2020; Dell, Jones, and Olken 2012; Kotz, Levermann, and Wenz 2022; Russ 2020; World Bank 2016). The impacts extend beyond the economy; droughts negatively affect health and education outcomes, particularly for poor households, thereby exacerbating global inequality (Zhang and Borja-Vega 2024).

Water storage, such as groundwater and soil moisture, ensures water availability when precipitation and surface water supplies decline. The large-scale decline in water storage weakens this crucial insurance mechanism against climate variability (Burke et al. 2023; Rodella, Zaveri, and Bertone 2023). Regions with limited water infrastructure and governance are particularly vulnerable, facing heightened risks of water-related conflicts and migration (Harari and Ferrara 2018; Henderson et al. 2017).

Migration has increasingly become a strategy for adaptation to droughts and water scarcity, particularly in regions where livelihoods depend heavily on agriculture. Prolonged dry conditions reduce crop yields and exacerbate food insecurity, forcing communities to relocate in search of better opportunities (Kubik and Maurel 2023; Zaveri et al. 2021). Clement et al. (2021) project that climate-induced water stress could displace millions by 2050.

Despite growing evidence of the economic and social impacts of water scarcity, several critical gaps remain. First, empirical evidence for the effect of water availability on labor markets, specifically on how water shortages affect job opportunities, is limited. Second, there is a lack of clarity on the long-term impact of water scarcity and to what extent this impact can be mitigated through adaptation. Third, most studies focus on the localized impacts of water scarcity, overlooking broader general equilibrium effects, such as how local water crises affect global economies through trade links. Finally, although the environmental consequences of drying, such as loss of aquatic habitats and disruptions to ecosystems, have been documented, rigorous empirical analysis linking water scarcity to environmental damages, including the incidence of wildfires and their implications for biodiversity, is lacking.

This chapter addresses these gaps to provide a more comprehensive view of the consequences of continental drying. It further examines the heterogeneous effects of drying across populations and regions, shedding light on how varying levels of water dependence, infrastructure, and institutions influence the severity of these impacts.

Findings from this chapter reveal that water scarcity significantly reduces job opportunities, particularly in rural communities reliant on agricultural activities. In Sub-Saharan Africa, extreme dry conditions have rendered 600,000–900,000 individuals jobless annually. Although the primary impact is reduced agricultural productivity, manufacturing and service industries are not insulated from these shocks. Moreover, the effects of local water shortages often transcend national borders through interconnected trade networks, turning local water scarcity and water mismanagement into a collective global risk. Beyond economic impacts, continental drying intensifies environmental risks by exacerbating the frequency and severity of wildfires. These wildfires not only threaten human settlements and livelihoods but also pose a severe risk to global biodiversity.

Two additional impacts of continental drying, although not included in the detailed analysis, are worth mentioning: sea level rise and glacier melting. As described in chapter 1, continental drying contributes to sea level rise. Sea level rise poses several risks, including coastal erosion, increased flooding, and saltwater intrusion into freshwater systems, which can affect agriculture and drinking water supplies. It also threatens coastal communities by damaging infrastructure and displacing populations.

Approximately 1.9 billion people worldwide depend on mountain glaciers (or their outflows) as their primary water source (Immerzeel et al. 2020; Immerzeel, Van Beek, and Bierkens 2010). Glacier outflows are also important for hydropower generation, which is often the main source of electricity in mountainous regions. As glaciers melt, they initially increase streamflow until peak water is reached, after which streamflow decreases and becomes less reliable. By 2018, about half of glacier-fed basins had already reached peak water (Rounce et al. 2023). Glacier melting also increases the risk of glacier lake outburst floods, affecting mountain communities.

Water and jobs

Water availability influences labor market outcomes through both supply and demand channels. On the supply side, water scarcity affects labor productivity by harming nutrition and health (Alsan and Goldin 2019; Hunter, MacDonald, and Carter 2010; Pongou, Ezzati, and Salomon 2006;

Prüss-Üstün, Bonjour, and Corvalán 2008). Studies show that droughts affect children's human capital development, leading to lower productivity in adulthood (Damania et al. 2017; Hyland and Russ 2019; Maccini and Yang 2009). When water is scarce, people tend to use polluted or inadequate water resources or limit hygiene practices. Water shortages due to drought or inadequate infrastructure increase the risk of various illnesses such as cholera (Emran et al. 2024). During water shortages, immediate productivity losses can occur if people spend productive hours fetching water or caring for those who are sick (Koolwal and Van de Walle 2013).

On the demand side of the labor market, water shortages negatively affect productivity in water-dependent sectors such as agriculture, energy, manufacturing, tourism, and transportation. This decline in productivity lowers job demand—at least in the short term—and negatively affects income. In the global workforce, about 78 percent of jobs have some level of water dependence, and 42 percent of jobs are significantly water dependent (Das, Fisiy, and Kyte 2013).

Water's role in sustaining jobs and livelihoods is even more pronounced in developing countries, where a large share of the population depends on agriculture. About 80 percent of the world's poor live in rural areas and rely heavily on farming, which requires reliable water access for crop cultivation. Globally, 1.23 billion people are directly employed in agriculture. The dependence is particularly stark in regions such as South Asia and Sub-Saharan Africa, where more than 50 percent of the workforce is engaged in agricultural activities.

The analysis for this report explores the effect of water scarcity on job outcomes in Sub-Saharan Africa. Although Sub-Saharan Africa has not experienced the widespread drying trends highlighted in the first chapter, it has faced frequent dry events over the past two decades (refer to map 2.1). With two-thirds of Sub-Saharan Africa's population reliant on rain-fed agriculture for their livelihoods, job security in the region's farming sector is particularly vulnerable to water scarcity.

From wilted fields to waning work

This analysis draws on 55 rounds of georeferenced household survey data, covering 7.5 million individuals across 30 countries in Sub-Saharan Africa from 2005 to 2018, to estimate the impact of dry events on an individual's probability of employment. A *dry event* is defined as a decline in soil moisture, which is an important component of terrestrial water storage (TWS) and critical for agricultural yields. Soil moisture is affected by both

Sub-Saharan Africa has faced frequent dry events over the past two decades.

MAP 2.1 Distribution of dry shocks, Sub-Saharan Africa, 2005–18

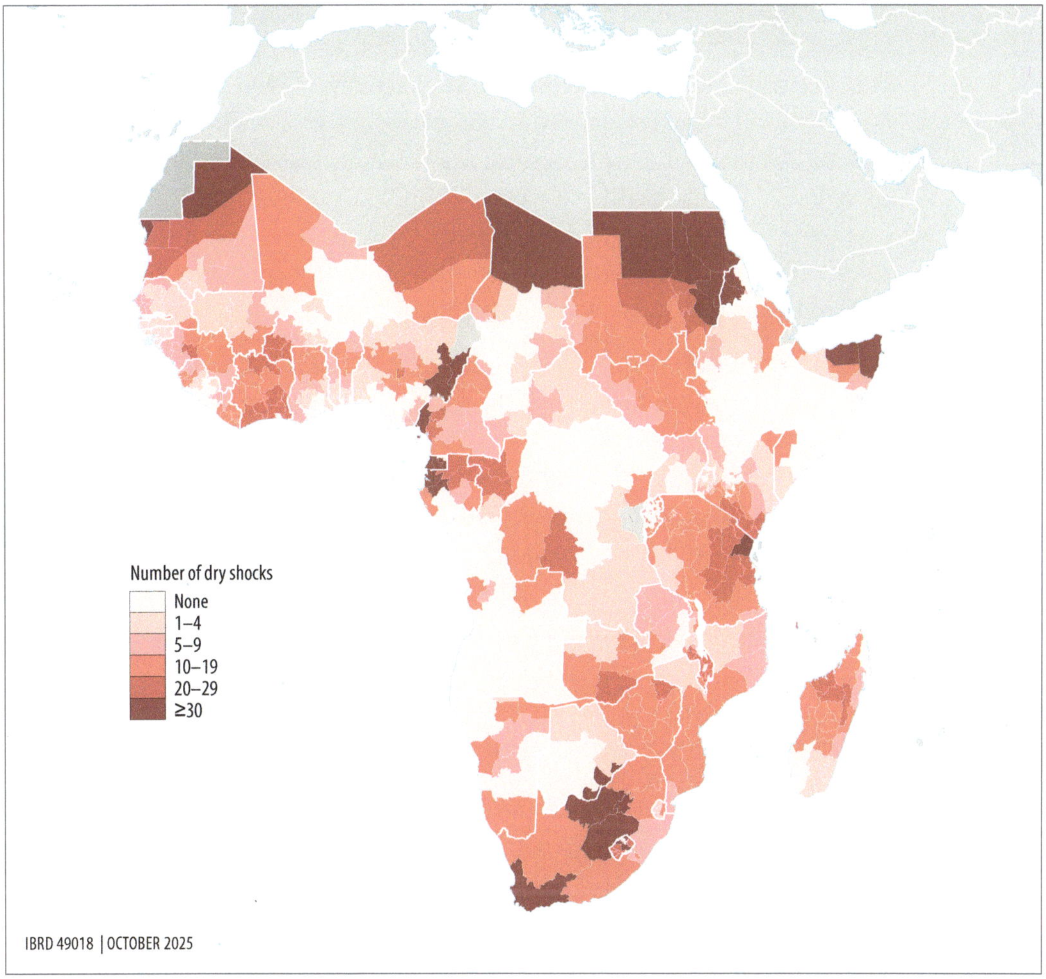

Source: World Bank.
Note: A location is exposed to a dry shock in a given month if the monthly soil moisture proxied by the Standardized Precipitation-Evapotranspiration Index in this location is 1.5 standard deviations below the long-term mean in the same location for that month.

rainfall and temperature, and determines how much water is available for crop growth. The Standardized Precipitation-Evapotranspiration Index (SPEI)—which was developed by Vicente-Serrano, Beguería, and López-Moreno (2010) and captures the joint effects of rainfall, temperature, and potential evapotranspiration—is used as a proxy for soil moisture (Lu et al. 2025; Wang, Rogers, and Munroe 2015).

The results of the analysis show that a dry shock—defined as an SPEI 1.5 standard deviations below the long-term mean[1]—reduces the employment rate by 2.5 percentage points on average (refer to figure 2.1). This decrease is driven by impacts concentrated in agriculturally dependent rural areas, where dry shocks reduce the employment rate by 7.4 percentage points. This relationship remains robust after controlling for individuals' gender, age, education, and place of residence, as well as seasonality, location, and common year trends across all regions. Between 2005 and 2018, approximately 600,000–900,000 individuals in Sub-Saharan Africa were made jobless each year because of exposure to dry shocks. This range represents a loss of 7–9 percent of the annual jobs created in the region because of the impact of water scarcity.[2]

The impact of water scarcity depends on a region's baseline climate conditions. Dividing the sample into different climate zones based on long-term average precipitation and temperature shows that employment in dry and hot regions is most sensitive to water scarcity.[3] A 1-standard-deviation decrease in SPEI reduces the likelihood of employment by 2.6 percentage points in these areas. Moderately dry and moderately hot

Dry shocks reduce employment, especially in rural farming communities.

FIGURE 2.1 Impacts of water scarcity on employment, average and rural farming, Sub-Saharan Africa, 2005–18

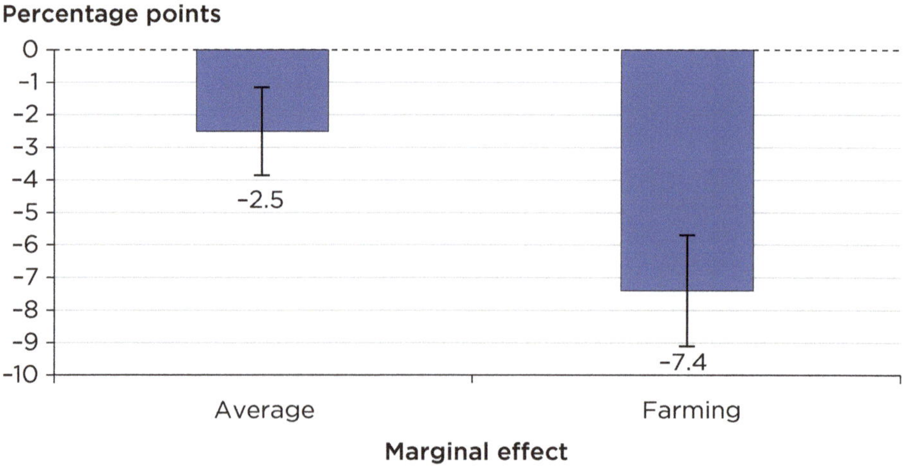

Source: Khan et al. 2024.
Note: The figure depicts the marginal effect of a 1.5-standard-deviation decline in the Standardized Precipitation-Evapotranspiration Index on the probability of having a job. The error bar represents the 95 percent confidence interval.

regions experience a 1.6 percentage point reduction; no statistically significant impact is observed in mildly dry or wet regions, regardless of their long-term average temperatures.

There is also considerable variation in the impacts of water availability on employment on the basis of age, gender, location, farmers' land ownership, and workers' skill levels. The negative effect of water scarcity on employment is most pronounced among women, older individuals, landless farmers, and low-skilled workers (refer to figure 2.2). Several factors may explain why these demographic groups are disproportionately affected by droughts. Droughts can worsen health conditions, particularly among older individuals; women are more likely to assume caregiving responsibilities for sick household members during these periods. Additionally, landless farmers and low-skilled laborers face significantly higher barriers to transitioning into nonfarm employment because of limited resources and a lack of specialized skills.

Droughts most negatively affect jobs for women, older individuals, landless farmers, and low-skilled workers.

FIGURE 2.2 Heterogeneous drought effects on employment, by individual worker characteristics

Source: Khan et al. 2024.
Note: The figure depicts the marginal effect of a 1.5-standard-deviation decline in the Standardized Precipitation-Evapotranspiration Index on the probability of having a job. The error bar represents the 95 percent confidence interval.

These results reflect short-term partial equilibrium effects of water scarcity on employment. Long-term general equilibrium effects are more complex. In response to prolonged droughts, local populations may expand farmland to offset productivity losses (Damania et al. 2017), potentially increasing agricultural employment. In low-income countries with trade barriers, declining agricultural productivity may push workers further into subsistence farming, deepening dependence on agriculture (Nath 2024).

Employment outcomes are not just about numbers; the nature of jobs matters. A rise in agricultural employment in water-stressed economies is not necessarily beneficial. As discussed, water scarcity reduces agricultural productivity, lowering farmers' real incomes. Moreover, evidence shows that, in response to declining water availability, households send children to work to compensate for income loss, undermining human capital accumulation and future earnings (Blakeslee, Fishman, and Srinivasan 2020).

These dynamics underscore the need for policies that address both immediate employment shifts and long-term structural impacts. Strengthening water management, improving agricultural resilience, and expanding nonagricultural job opportunities will be critical in mitigating the adverse labor market effects of water scarcity. The next section discusses potential policy responses and adaptation strategies.

Infrastructure, institution, and resilience

Economic and infrastructure development play an important role in determining the impact of water shocks. Investment in water supply infrastructure, such as storage and irrigation, can reduce the impact of water shortages by transporting water from distant sources (McDonald et al. 2014) and reducing agriculture's exposure to local climate variability (Schlenker, Haneman, and Fisher 2005). Irrigation significantly reduces the sensitivity of crop yields to drought, consequently improving agricultural production, employment, and income stability for farmers (Briscoe and Malik 2006; Pingali 2012).

However, without proper water use regulations, supply-side interventions often risk moral hazard issues. The perceived security provided by increased water supply, combined with distortions in the input and output market of agriculture, can lead to more water-intensive and less efficient agricultural practices.[4] This phenomenon has been observed in the Ogallala Aquifer in the United States, where large-scale groundwater-based irrigation initially mitigated the impact of droughts.

Over time, however, land use shifted to more water-intensive crops, increasing drought sensitivity and reducing land value (Hornbeck and Keskin 2014). On a global scale, Damania et al. (2017) found that although irrigation generally reduces crop yield sensitivity to droughts, in some arid regions and low-income countries, large irrigation infrastructure may exacerbate the impact of droughts on agricultural yields.

The analysis presented in this report reveals similar trends. In areas with irrigation infrastructure, employment is found to be more vulnerable to the negative impacts of drought. In these areas, a dry shock is associated with a 3.2 percent reduction in the likelihood of employment, compared with a 2.5 percent reduction on average. In addition to the shift to more water-intensive crops, the agricultural productivity gains enabled by irrigation may have attracted a population influx due to increased agricultural opportunities, consequently increasing job exposure to climate variability without driving structural transformation (Asher et al. 2021; Boudot-Reddy and Butler 2025).

These findings suggest that there is often a trade-off between short-term profitability and long-term sustainability in agricultural jobs that rely on groundwater-based irrigation, particularly in water-stressed regions. Effective pricing incentives and robust water regulations are crucial to minimize this trade-off.

Beyond interventions specific to the water sector, economic development and diversification are essential strategies to adapt to droughts and reduce their adverse effects on jobs. To explore the role of economic diversification in adaptation, analysis for this report analyzes the heterogeneous effects of water scarcity based on communities' historical livelihood strategies. Specifically, data from the ethnographic atlas originally compiled by Murdock (1967) are used to determine whether individuals reside in regions where ethnic groups historically depended on agriculture; hunting, fishing, and gathering; or herding and animal husbandry (or pastoralism) during the preindustrial era.

Societies with different historical subsistence strategies evolved differently and achieved different levels of economic diversification over time (Agrawal 2008). Specifically, studies found that early agricultural societies progressed to civilization and statehood sooner and are wealthier and more educated today than pastoral ones (Diamond 1997; Galor and Özak 2016; Giuliano and Nunn 2021; Hibbs and Olsson 2004; Michalopoulos, Putterman, and Weil 2019; Putterman 2008). Additionally, preindustrial fishing activities led to nonagricultural occupations and skills that complemented early industrialization (Dalgaard, Knudsen, and Selaya 2020).

The analysis in this report supports these findings. Societies with a historical reliance on hunting, gathering, and fishing show the highest levels of economic diversification as measured by the Herfindahl-Hirschman Index, with more occupations in services, sales, and domestic jobs. In contrast, historically agricultural societies continue to depend more heavily on agricultural employment. Pastoral societies exhibit the least diversified economies, with the highest concentration of jobs in the agricultural sector and the lowest income and education levels among the three groups.

Unsurprisingly, these ethnic groups exhibit different adaptive capacities in response to dry shocks. Regions inhabited by ethnic groups that historically engaged in hunting, fishing, and gathering did not experience statistically significant adverse effects from dry shocks. Areas historically associated with agriculture-focused ethnic groups are negatively affected by dry shocks, although these effects are less severe than those faced by pastoral ethnic groups, which experience the most significant consequences. These findings support the notion that communities heavily reliant on water-intensive sectors, such as agriculture, are more susceptible to disruptions caused by drought. Diversification into less water-dependent sectors can mitigate these vulnerabilities.

This finding aligns with evidence from India, where research shows that households in regions with a more developed manufacturing sector are better able to offset farm losses from long-term water shortages by increasing off-farm income (Blakeslee, Fishman, and Srinivasan 2020). Improved access to markets and financial services, along with better road infrastructure, plays a crucial role in developing the rural nonfarm economy, strengthening job opportunities, and enhancing income resilience for rural households (Musungu, Kubik, and Qaim 2023; Nguyen, Nguyen, and Grote 2023).

Water, food, and income

Agriculture is the most water-intensive sector, consuming more than 90 percent of global freshwater resources. As highlighted in the previous section, agriculture serves as the primary channel through which water scarcity affects employment and the economy. Agricultural productivity is highly sensitive to rainfall variability (Damania et al. 2017). Extended periods of low soil moisture can lead to crop stress, reduced yields, and even crop failure (Agnolucci et al. 2020). Drought-related food crises have been observed globally, affecting both developed and developing countries (FSIN 2018).

This section revisits the link between water scarcity and agricultural productivity, with a particular focus on the long-term effects of continental drying and the effectiveness of adaptation measures in mitigating these impacts. The second part of the section investigates how local water shortages influence agricultural production and create spillover effects on real income at the global level. Although the global community is interconnected through the hydrological cycle—via transboundary water flows and atmospheric water vapor movement (GCEW 2024)—it is also intertwined through trade and production networks. The analysis highlights that, because of these economic interdependencies, water management decisions in one country can ripple across borders, with global economic implications.

Drying, warming, and tipping point

Understanding the long-term effects of drying is not straightforward. Many studies estimate the impact of water shocks by examining short-term fluctuations in rainfall because rainfall changes are largely random and independent of other factors, allowing for clearer identification of causal relationships. However, this approach may not fully capture the effects of long-term adaptation. In anticipation of long-term decline in freshwater availability, farmers may implement a variety of adaptation strategies, such as adopting drought-resistant crops or relocating from areas frequently affected by dry conditions. These long-term adjustments are less likely to be captured in studies focused solely on short-term rainfall variability (Carleton and Hsiang 2016; Hsiang 2016).

To measure potential adaptation, the analysis follows the approach of Deschênes and Greenstone (2011) and Carleton et al. (2022) to estimate the marginal impact of declining water availability on agricultural yields while allowing the impact to vary with a location's baseline climate conditions. The underlying premise is that baseline climate conditions shape the perceived likelihood of and belief in future droughts and, consequently, influence investment and adaptation behaviors. Adaptation is more likely to occur in consistently dry environments, where water scarcity is a persistent challenge. By comparing the water sensitivity of agricultural yields across different locations, this approach allows the assessment of adaptation linked to baseline climate conditions.

The analysis delineates a total of 49 climate zones globally, based on a location's long-run average temperature and precipitation from 1951 to 2022. The climate zones are categorized using 5°C bins for temperature and 200 mm bins for precipitation. The impact of water availability on agricultural yields is estimated at grid-cell level at 0.5° × 0.5° and aggregated up to the climate zone level. Water availability is measured by both precipitation and SPEI, with SPEI serving as a proxy for soil moisture.

Local vegetation growth is proxied by net primary production (NPP) data collected using satellite observations. NPP reflects the net amount of carbon stored by plants through photosynthesis, serving as a proxy for biomass production, and it has been used in various studies to assess farmland productivity (Haberl et al. 2007; Jaafar and Ahmad 2015; Monfreda, Ramankutty, and Foley 2008).

Another feature of this analysis is that it examines how variations in water availability, along with temperature fluctuations, jointly influence agricultural outcomes, thereby filling a gap in the existing literature, which typically considers the effects of either temperature or precipitation independently, without exploring their interactive influence. For further details on the data and methodology, refer to the online technical appendixes.[5]

Table 2.1 presents the estimated relationships between climate variables and vegetation outputs, showing a significant heterogeneity in the impact of precipitation and temperature changes across climate zones. The results yield three key findings.

First, at the aggregate level, a 1-standard-deviation decrease in SPEI and precipitation results in an average reduction in vegetation output of approximately 9.6 percent and 6.0 percent, respectively. This magnitude is comparable to the effect of water shocks reported in Damania et al. (2017).[6]

Second, the findings suggest that drought-prone regions are particularly vulnerable to declines in water availability, highlighting the limited effectiveness of adaptation on a global scale.

When temperature is held constant, the marginal effect of reduced precipitation is markedly higher in drier regions. This effect diminishes in less arid areas and becomes statistically insignificant in humid regions where water is already abundant.[7] The pattern remains consistent when using SPEI to measure drought. A 1-standard-deviation decrease in SPEI in dry regions with baseline precipitation below 600 mm results in an 11.9 percent reduction in vegetation growth, compared with a 4.0 percent reduction in regions with a baseline precipitation of 600 mm or more.

The observation that drought-prone areas are more sensitive to reductions in water availability indicates that adaptation may have limited effectiveness globally. This finding is aligned with previously reported results showing that long-term agricultural adjustments in response to worsening environmental conditions have been limited. Notable examples include the limited adaptation observed after the 1930s Dust Bowl in the United States (Hornbeck 2012) and prolonged groundwater depletion in India (Blakeslee, Fishman, and Srinivasan 2020; Fishman 2018). Key barriers to effective adaptation in agriculture identified in the literature include restricted access to capital, small farm sizes, insecure land tenure, lack of infrastructure, weak institutional support, and risk aversion (Adger et al. 2009; Below et al. 2012; Core Writing Team, Lee, and Romero 2022; Hornbeck 2012; Morton 2007).

Third, the results suggest mutually reinforcing adverse effects of drying and warming.[8] Drying makes adapting to warming more difficult, whereas warming exacerbates the negative effects of drying. Moreover, these negative effects are nonlinear, implying that the cumulative effects of drying and warming can significantly intensify climate change impacts on food production in the long run.

TABLE 2.1 Impact of climate variation on vegetation outputs (percent)

Precipitation (mm)	Temperature (°C)						
	<0	**[0, 5)**	**[5, 10)**	**[10, 15)**	**[15, 20)**	**[20, 25)**	**≥25**
			a. 10 mm decrease in annual precipitation				
[0,200)	0.64[a]	−2.26[a]	−2.98[a]	−2.97	−2.91	−3.30	−9.25
[200, 400)	0.33[a]	−1.16	−1.08	−1.24	−1.25	−2.58	−6.71
[400, 600)	0.43	0.18	−0.20	−0.49	−0.59	−0.75	−3.49
[600, 800)	0.33[a]	0.20[a]	−0.11	−0.22	−0.37	−0.37	−1.67
[800, 1,000)	0.22[a]	0.21[a]	0.08	−0.12	−0.30	−0.19	−0.45
[1,000, 1,200)	0.20[a]	0.21[a]	0.08[a]	−0.09	−0.28	−0.22	−0.23
≥1,200	*0.03*[a]	0.15[a]	*0.00*[a]	*−0.00*[a]	*0.01*	*0.00*	0.06
				b. 1 SD decline in the SPEI			
[0,200)	5.02[a]	−6.59[a]	−10.79[a]	−13.17	−14.32	−30.33	−56.02
[200, 400)	2.59	−6.95	−8.09	−10.10	−12.11	−43.57	−96.69

(continued)

TABLE 2.1 Impact of climate variation on vegetation outputs (percent) *(continued)*

Precipitation (mm)	Temperature (°C)						
	<0	**[0, 5)**	**[5, 10)**	**[10, 15)**	**[15, 20)**	**[20, 25)**	**≥25**
[400, 600)	3.23	1.23	−3.01	−6.09	−9.29	−15.06	−70.09
[600, 800)	3.31[a]	2.92[a]	−2.17	−4.03	−6.59	−6.44	−40.48
[800, 1,000)	1.01[a]	2.28[a]	*−0.01*	−3.53	−5.25	−3.65	−13.25
[1,000, 1,200)	4.67[a]	3.38[a]	0.93[a]	−2.82	−5.19	−4.13	−7.20
≥1,200	*1.22*[a]	4.79[a]	1.36[a]	*−0.22*[a]	*−0.44*	−1.45	*−0.31*
c. 0.1oC increase in annual average temperature							
[0,200)	0.61[a]	0.24[a]	*0.05*[a]	−0.20	−0.92	−5.27	−12.14
[200, 400)	0.23[a]	0.30	−0.31	−0.54	−1.35	−4.97	−14.43
[400, 600)	0.30	*0.00*	−0.38	−0.66	−1.43	−2.78	−7.50
[600, 800)	0.42[a]	0.41[a]	−0.25	−0.57	−1.06	−1.56	−3.83
[800, 1,000)	0.52[a]	0.12[a]	−0.44	−0.71	−0.53	−1.03	−2.44
[1,000, 1,200)	1.07[a]	0.37[a]	−0.38[a]	−0.78	−0.67	−0.99	−1.44
≥1,200	1.56[a]	1.10	0.34[a]	−0.74[a]	−0.52	−1.26	−1.47

Source: World Bank.
Note: Panel a presents the average marginal effects of a 10 mm decrease in annual total precipitation below the long-run average on (log) NPP for cells within each climate zone. Panel b presents the average marginal effects of a one-unit (that is, 1-standard-deviation) change in the SPEI on (log) NPP for cells within each climate zone. Panel c presents the average marginal effects of a 0.1°C increase in annual average temperature on (log) NPP for cells within each climate zone. Climate zones are based on the long-run average temperature and precipitation of each cell between 1951 and 2022, categorized using 5°C bins for temperature and 200 mm for precipitation. Climate bins are defined with inclusive lower bounds and exclusive upper bounds (e.g., [5, 10) includes 5 but not 10). The regression is weighted by cell size and jointly estimated with heterogeneous responses to precipitation, SPEI, and temperature. The regression includes cell fixed effects and year fixed effects, and standard errors are clustered at the cell level. All estimates except those in italics are statistically significant at the 1 percent level. NPP = net primary production; SD = standard deviation; SPEI = Standardized Precipitation-Evapotranspiration Index.
a. Estimated from a climate zone covering less than 0.25 percent of global cropland.

The degree of dryness can affect the marginal impact of higher temperatures. As shown in figure 2.3a, in warmer regions, as baseline precipitation decreases, even a small decrease in precipitation leads to exponentially greater losses in crop yields.

Similarly, as shown in figure 2.3b, in the drier half of the world, the adverse effects of warming on crop yields would double if the region's temperatures were above the threshold of 20°C and would quintuple if temperatures were above 25°C.

Both table 2.1 and figure 2.3 illustrate that the joint effects of drying and warming are nonlinear, implying that a cumulative increase in drying and warming over time could result in significantly large economic damages. This finding is supported by evidence from agricultural physiological studies and observations from recent hot-dry events (refer to box 2.1). In some of the world's hottest and driest regions, a 10 mm decrease in annual precipitation or a 1-standard-deviation decline in soil moisture can lead to a 9.3 percent or a 43.1 percent reduction in crop productivity, respectively. This finding suggests that higher temperatures will compound the adverse effects of water scarcity, and prolonged dry conditions will make adapting to warmer temperatures more challenging.

Negative effects of drying and warming on crop yields are mutually reinforcing.

FIGURE 2.3 **Marginal effects of drying and warming on vegetation yields, by climate zone**

a. Effect of 10 mm decrease in precipitation on vegetation growth

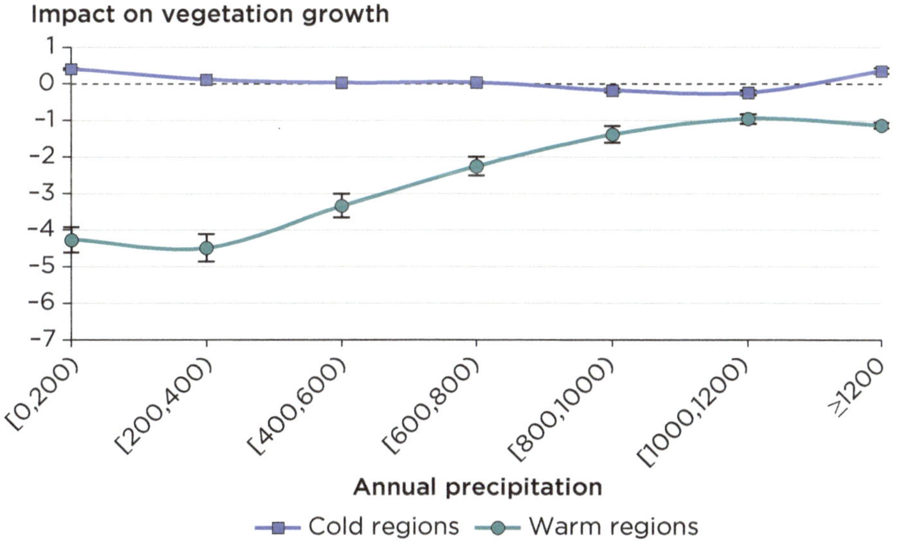

(continued)

FIGURE 2.3 Marginal effects of drying and warming on vegetation yields, by climate zone *(continued)*

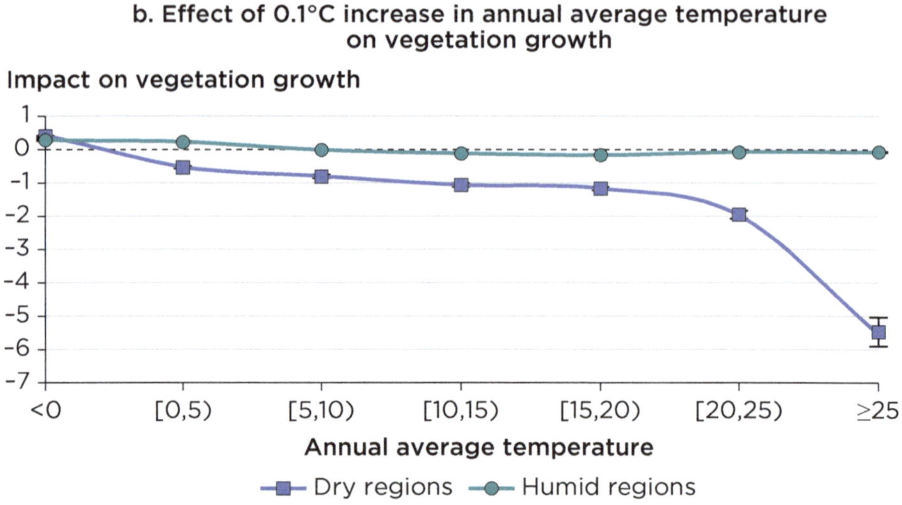

Source: World Bank.

Note: Panel a presents the average marginal effects of a 10 mm decrease in annual total precipitation below the long-term average on (log) NPP—a proxy for vegetation output—for cells within each climate zone. Panel b depicts the average marginal effects of a 0.1°C increase in annual average temperature above the long-term average on (log) NPP for cells within each climate zone. Climate zones are defined by the long-run average temperature and precipitation of each cell over the period 1951–2022, categorized using 5°C bins for temperature and 200 mm for precipitation. Climate bins are defined with inclusive lower bounds and exclusive upper bounds (e.g., [5, 10) includes 5 but not 10). The global long-term median temperature (16.8°C) serves as the threshold between warm and cold regions in panel a. The global long-run annual median precipitation (521 mm) serves as the threshold between dry and humid regions in panel b. The regression is weighted by cell size and jointly estimated with heterogeneous responses to temperature and precipitation. Error bars denote 5 percent confidence intervals. NPP = net primary product.

BOX 2.1

Heat and drought: The rising threat to global crop resilience

Studies indicate that simultaneous heat and drought conditions exacerbate crop stress far more than either factor alone. This dual impact reduces soil moisture and escalates temperature stress on plants, particularly during critical growth stages. For example, maize and soybeans are exceptionally

(continued)

BOX 2.1

Heat and drought: The rising threat to global crop resilience *(continued)*

vulnerable to high temperatures when grown in dry conditions—they are 40 percent more sensitive to high temperatures in these conditions than when they are well watered (Lesk et al. 2021).

In recent years, compounded hot-dry events globally have resulted in substantial agricultural shocks, underscoring the detrimental effects of combined heat and drought on crop production. For example, severe heat-drought events in the Russian Federation and Eastern Europe led to drastic yield reductions in wheat and barley (Hunt et al. 2021), and the intense drought and heat wave in the central United States in 2012 resulted in significant drops in maize and soybean production (USDA 2013). These climate extremes critically disrupt crop yields and elevate food prices, particularly in areas reliant on rain-fed agriculture.

Countries such as India are increasingly challenged by the compounding effects of higher temperatures and reduced rainfall. With much of India's agriculture situated in tropical to subtropical climates, key crops such as wheat and rice are nearing their thermal limits (Beillard and Singh 2022). With these escalating climate challenges, the need for resilient agricultural practices has never been greater. Adaptation through breeding heat-resistant crops and improving irrigation systems will be essential to ensuring food security and sustainable agricultural practices in the future.

Local water shocks, global economic impact

The Global Commission on the Economics of Water highlights the importance of valuing and governing the hydrological cycle as a global common good because of its interconnected nature (GCEW 2024). In addition to connections through the physical hydrological system, increased agricultural trade heightens water dependencies across countries.

The removal of tariff and nontariff trade barriers has driven significant growth in agricultural trade over recent decades. Between 2018 and 2020, the value of agricultural trade in crops increased by more than 80 percent compared with 2007, whereas manufacturing exports rose by less than 40 percent during the same period. Crops and food exports now represent more than 10 percent of global trade (FAO 2024).

Consequently, the water embedded in the production of globally traded goods and services, known as the virtual water trade (Allan 1998), increased by 26 percent between 2000 and 2019. In 2019, approximately

25 percent of global water consumption was allocated for export rather than domestic use (refer to chapter 3 for details). The large volume of virtual water trade implies that local water shortages affecting agricultural production can have spillover effects on a global scale.

A drought in a major agricultural exporting region could reduce agricultural productivity, leading to lower output and wages for local agricultural workers. The decline in productivity could increase prices for agricultural goods, affecting all consumers of these goods, regardless of their occupation. Additionally, it could raise the cost of inputs for manufacturing and services firms worldwide that use these goods in their production processes. These higher input costs could then result in increased prices for downstream goods and services, such as packaged food and restaurant services, which could subsequently affect downstream consumers.

In this context, the virtual water trade imposes two contrasting impacts on the economy. First, agricultural products serve as intermediate goods in other sectors of the economy, thereby increasing these sectors' exposure to changes in water availability. Consequently, the impacts of droughts have the potential to propagate through the global production network, affecting countries not directly experiencing drought. This interconnectedness heightens the vulnerability of various economic sectors to water shocks.

Second, changes in water availability can shift productivity and comparative advantage across producers. Trade patterns can then adjust to accommodate water shocks, allowing consumers to redirect expenditures and source food and water-intensive goods from unaffected regions. This adjustment mechanism helps to mitigate the adverse effects of water shocks on price and consumption.

This report develops a spatial equilibrium trade model, based on Eaton and Kortum (2002), to analyze the general equilibrium effects of water scarcity. By modeling the structure of the global economy, it examines how local water shortages affect real income in various economic sectors and how they propagate through different countries through existing trade networks. It simulates two outcomes. The first set of simulations evaluates the historical cost of climate anomaly worldwide in 2002. The second set assesses the welfare cost of a 100 mm reduction in precipitation for each country relative to the 2020 average, estimating the impact on real income within the country (local cost) and the global cost resulting from a single country's rainfall deficit. This analysis is grounded in the empirically estimated crop yield–water response function presented in the previous section. Refer to box 2.2 and technical appendix B (online) for details of the methodological framework and data.

BOX 2.2

Trade general equilibrium model

Following the recent literature on economic geography and the environment (Nath 2024; Rudik et al. 2022), this report develops a spatial multi-industry structural trade model in the style of Eaton and Kortum (2002) to simulate the effect of local water scarcity on global economy.

The model uses 26 industries in three sectors—agriculture, manufacturing, and services—and 172 countries. In each country-industry combination, firms operate a country-industry-specific constant returns-to-scale Cobb-Douglas technology that uses labor, capital, and materials to produce output. Materials are the output of other industries; for example, producing maize may require manufactured fertilizer.

Firm productivity in a country-industry combination is a function of factors such as local infrastructure, temperature, and precipitation. In this model, changes to precipitation have an impact on firm productivity, affecting the output that firms can produce with a given input combination. The productivity-precipitation response function is estimated on the basis of the empirical analysis described in the "Drying, warming, and tipping point" section and exhibits considerable heterogeneity across climate zones. The technical appendixes (online)[a] provide details on how the microlevel water shocks are mapped to country-level agricultural productivity.

The model styled after Eaton and Kortum (2002) emphasizes the role of comparative advantage and endogenizes trade flows on the basis of differences in productivity across sectors and regions. It is particularly strong in analyzing how changes in productivity—for example, due to water shocks—affect output, prices, trade, and welfare in a tractable way. However, like all quantitative structural models, the model relies on a set of assumptions about functional forms and the structure of the economy that may not fully capture real-world complexities, thereby influencing the robustness of its predictions.

For example, the model assumes each good is produced using a Cobb-Douglas production function. The Cobb-Douglas function assumes a constant elasticity of substitution of 1 between inputs, which implies that water can always be substituted for other inputs, such as labor or capital, at a constant rate. In many cases, water is not easily substitutable. Assuming constant substitutability can lead to underestimation of production responses to water scarcity.

(continued)

BOX 2.2
Trade general equilibrium model *(continued)*

Additionally, water is not explicitly treated as a direct input in production but is instead incorporated indirectly through its impact on productivity. Ideally, water consumption (as a combination of irrigation and rain-fed sources) should be included as an input in the production function, with producers optimizing on the basis of water prices and other input-output prices. However, in many regions, the absence of functioning water markets makes it difficult to observe how markets (and equilibrium water prices) respond to variations in water availability, such as droughts or rainfall. Without reliable data on these dynamics, directly modeling water as an input would introduce significant uncertainty.

Despite the model's simplifying assumptions, its projections of the impact of historical droughts in 2002 on prices and income in affected countries largely align with observed outcomes, supporting confidence in the model's external validity. This type of structural model has also been shown to capture the margins of heterogeneity and spatial differences—crucial for quantifying the distribution of climate change impacts (Rudik et al. 2022).

Consequently, although the model's simplifying assumptions may affect the precision of point estimates, the model remains a valuable tool for understanding the broader spillover effects of water shocks—that is, how reduced agricultural productivity in one region can ripple through supply chains and trade networks, influencing income distribution across sectors and locations.

a. The technical appendixes are available online at https://hdl.handle.net/10986/43683.

Impact of the 2002 global climate anomaly

In 2002, droughts affected multiple regions with varying severity. For example, India experienced one of its driest monsoon seasons, with rainfall 19 percent below average. Additionally, 2002 was one of the driest years on record for Australia, which was exacerbated by an El Niño event, reducing rainfall across the continent. Sub-Saharan African countries, including Ethiopia and Mozambique, all experienced major drought events.

Using the relationship between local water shocks and macrolevel agricultural productivity estimated from the crop yield–water response function reported earlier, it is estimated that the global climate anomaly of 2002 resulted in significant declines in agricultural productivity among several major producers. Notably, Pakistan experienced a reduction of approximately 19.6 percent. Other affected countries were India

(13.9 percent decrease), Paraguay (9.4 percent decrease), Australia (8.5 percent decrease), Brazil (7.0 percent decrease), Ethiopia (5.8 percent decrease), and Mozambique (5.5 percent decrease).

Precipitation anomalies significantly contributed to productivity declines. In 2002, 45–78 percent of agricultural output reductions in Australia and Pakistan were due to below-average rainfall. In Mozambique, these anomalies accounted for 67 percent of the decline. About 25 percent of the reductions in Ethiopia and Paraguay were linked to changes in rainfall patterns. The rest were mainly due to temperature anomalies.

Using the trade equilibrium model developed for this report, the impact of these climate anomalies is estimated to have caused a global welfare decline of 0.4 percent, resulting in significant economic losses. Using the global gross domestic product (GDP) from 2002 as a reference, this decline equates to approximately $140 billion in 2002 dollars, or $245 billion in constant 2024 dollars, which is about $22 per capita over the global population in 2002.

The agricultural sector experienced the largest impact, with a 0.65 percent decline in global welfare; the manufacturing and service sectors faced welfare losses of 0.27 percent and 0.21 percent, respectively. These declines were driven by sectoral spillovers through production links. The negative welfare shocks from the 2002 climate anomaly were particularly severe in South Asia and Sub-Saharan Africa, where climate-induced agricultural productivity losses had widespread economic consequences. The most direct impact was a reduction in agricultural output due to climate disruptions that hindered production.

Farmers' short-term adaptation efforts, such as adjusting inputs, helped mitigate the impact of the climate anomaly but could not fully offset the productivity shock. Consequently, agricultural output declined by 4.4 percent in Pakistan. Without adaptation measures, the declines would have mirrored the full productivity shock (19.6 percent), resulting in even more significant reductions in output. In Sub-Saharan Africa, droughts led to a decline of up to 15 percent in agricultural output, with other major agricultural regions, including Australia, the Arab Republic of Egypt, and Ukraine, also experiencing reductions in production. These losses directly affected the welfare of agricultural workers, whose income is closely tied to the sector's performance.

The decrease in agricultural production led to a substantial increase in food prices. For example, the model estimates a 6.1 percent rise in food prices in Pakistan because of the climate shocks. This surge in food prices had cascading effects across the economy, particularly in nonagricultural sectors that rely on agricultural inputs, such as food processing and textile manufacturing, which resulted in higher prices for manufactured goods. These price hikes also placed significant strain on household budgets, reducing the purchasing power of workers' wages, regardless of their sector of employment. Consequently, the model predicts national welfare losses of 1.9 percent for Pakistan measured by real income. Agricultural workers were disproportionately affected, with much larger welfare losses of 4.3 percent.

The welfare of populations in countries not directly affected by the 2002 climate anomalies was influenced through indirect channels, such as international trade. The welfare outcomes in these countries were shaped by production links, access to global markets, trade costs, and the degree of substitutability and complementarity between their outputs and those of the affected regions. Countries that experienced more favorable or less severe climate conditions in 2002 saw their agricultural sectors benefit from higher global food prices, which helped stabilize global food supply chains. The model indicates that countries such as Argentina, Kazakhstan, Myanmar, and Spain experienced positive welfare shocks, with agricultural welfare increases of up to 4 percent. These welfare gains were driven by both local productivity improvements and substitution effects as the demand for their agricultural products increased in the international market because of production declines in drought-affected regions.

More important, the model suggests that trade likely played a crucial role in mitigating the global impact of the 2002 climate anomalies. Without market adjustments at both local and international levels, agricultural outputs in drought-affected countries such as Pakistan would have declined in line with its two-digit productivity losses. Trade facilitates price adjustments driven by global supply and demand, signaling where resources are most needed. Specifically, rising prices for water-intensive crops in drought-stricken regions incentivize other countries to increase production and exports of these crops, helping to meet global demand. This market response helps stabilize their prices and ensures a steady supply. Consequently, drought-affected countries in Sub-Saharan Africa and South Asia benefited significantly from international trade, with the loss of welfare reduced by more than 10 percent as measured in the real wages of their agricultural workers.

The significant benefit highlights the crucial role of trade in mitigating the local effects of decreased agricultural production in drought-affected nations. Countries can improve their resilience to climate-related shocks by establishing strategic trade partnerships with regions that are less susceptible to similar climatic risks. Additionally, investing in infrastructure and logistics to facilitate efficient import and export activities is essential for ensuring the smooth adjustment of trade relationships.

Local and global cost of water scarcity

The second set of outcomes sheds light on the cost of water scarcity at both local and global levels. Consider a hypothetical scenario involving a 100 mm decrease in precipitation in a country. The cost of an annual 100 mm decline in precipitation varies significantly, depending on the country affected and the way the shock propagates through the global production network.

Based on the results of the spatial equilibrium trade model described in box 2.2, India is the country where a hypothetical 100 mm reduction in precipitation would generate the largest global impact. Using 2020 data, such a reduction could lower global welfare by 0.08 percent—equivalent to $68 billion.[9] This outcome for India reflects several key factors incorporated into the model:

- A large portion of India's economy and workforce depends on agriculture, which is highly sensitive to water availability.
- Because India is a major producer, local water shocks in India have significant ripple effects through global trade, affecting prices and supply chains worldwide.
- Many of India's key agricultural regions already face significant water stress, so any further scarcity tends to carry a disproportionately high economic impact.

Other countries in which precipitation is globally important are China, Nigeria, and Pakistan because of their extensive agricultural activities, importance in the global production network, and/or high baseline water stress (refer to map 2.2).

The local cost of a 100 mm precipitation decline is also highly heterogeneous across different countries. Unlike the global impact, the highest local value of water is observed in African nations, where hot and arid conditions intensify the effects of water scarcity. Incorporating the input-output links within a country considerably increases the local value of water, especially in Sub-Saharan Africa, where agricultural activities play a crucial role in the economy.

Water, wildfires, and biodiversity

The National Aeronautics and Space Administration's mantra when searching for extraterrestrial life has long been "follow the water," underscoring the significance of water as a universal indicator of habitability. Water is essential for sustaining life and maintaining a livable planet. It is the cornerstone of all ecosystems, supporting the survival of every species.

Local water shortages trigger economic consequences both locally and globally.

MAP 2.2 Impact of a 100 mm reduction in annual precipitation on real income

a. Global cost

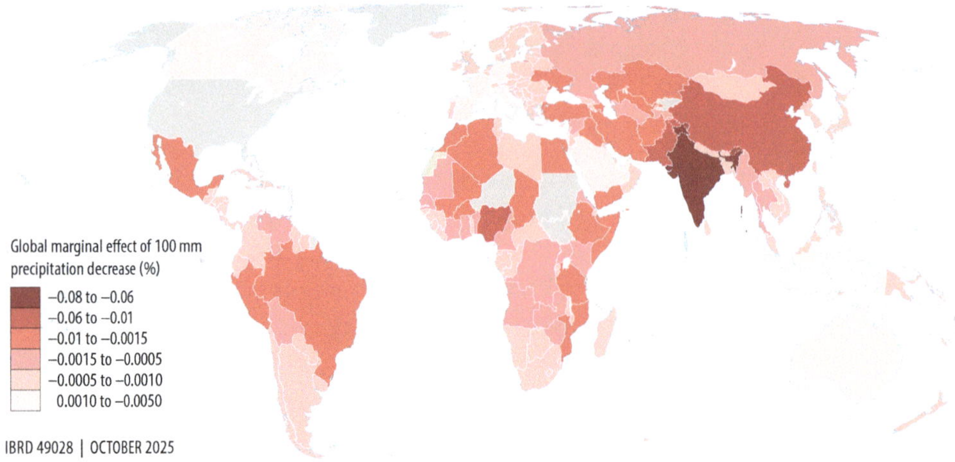

b. Local cost

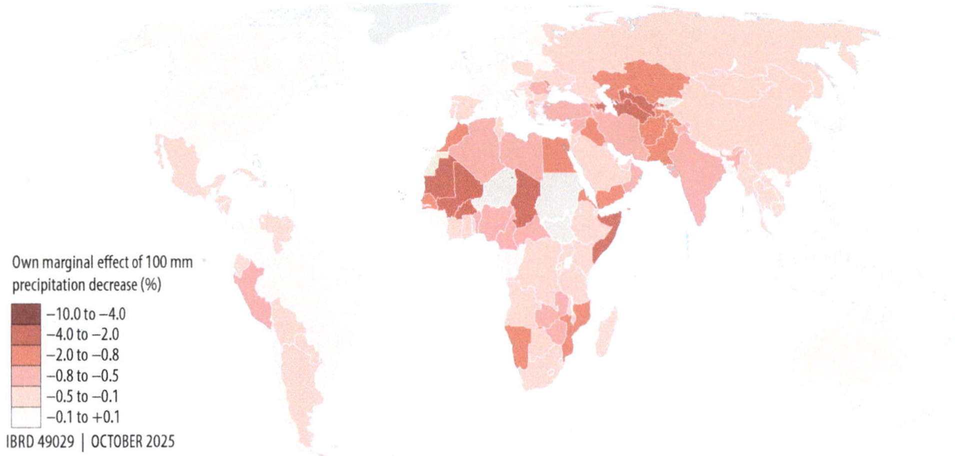

Source: World Bank.
Note: Panel a shows the impact of a 100 mm reduction in a given country's annual precipitation on global real income (that is, global cost). Panel b depicts the impact of the same reduction on the real income of the affected country (that is, local cost).

This section examines the effects of continental drying on ecosystems and biodiversity, particularly through its influence on the frequency and severity of wildfires. Although wildfires are often a naturally occurring phenomenon, they can be catastrophic for ecosystems and nearby human communities. The total annual cost of wildfires in the United States is estimated to range between 1 and 3 percent of GDP (World Bank 2023). A study focusing on Southern Europe has shown that wildfires can lead to a 0.11- to 0.18-percentage-point reduction in GDP growth, as well as to an offset in employment growth (Meier et al. 2023). In Indonesia, one of the countries vulnerable to large wildfires, the economic cost of the six largest fire events from 2004 to 2015 was estimated at about $94 billion (Kiely et al. 2021).

Recent years have seen an increase in wildfire occurrences and in their intensity, as measured by the size of burnt areas and duration (Ellis et al. 2022; Williams et al. 2019). In the United States, the number of large fires has increased by more than 30 percent since the 1980s, with California alone experiencing an 85 percent increase in annual burned areas since the 1970s.[10] Globally, the length of the fire season has expanded by nearly 20 percent over the past few decades (Doerr and Santín 2016; Jolly et al. 2015). These fires not only result in substantial economic losses but also contribute to biodiversity loss, soil degradation, and significant carbon emissions, further fueling climate change (Braun et al. 2021; Coop et al. 2020; Wang et al. 2020).

Many studies attribute the increase in wildfire events to climate change–related factors such as fuel aridity, rising temperatures, and changing rainfall patterns (Jones et al. 2022; Law et al. 2025). These factors, both individually and collectively, create conditions conducive to the ignition and spread of wildfires.

The analysis for this report indicates that, in addition to established weather controls such as the Fire Weather Index, which considers temperature, humidity, wind speed, and 24-hour precipitation, continental drying—that is, a decline in TWS—has a direct influence on the occurrence and severity of wildfire events at the global level.

Parched landscapes, intensified wildfires

Analysis for this report explores the relationship between TWS and wildfires, using the Moderate Resolution Imaging Spectroradiometer (MODIS) Burned Area Dataset (Giglio et al. 2021), as referenced in both the scientific and the economic literature, to identify areas affected by wildfires. To distinguish natural wildfires from built-environment fires and human-caused fires, MODIS Land Cover data (Friedl and Sulla-Menashe 2022) are used to exclude land associated with croplands, barren areas, and urban or built-up zones. Technical appendix B (online) provides more details on the data and methodology.

Results from the analysis indicate that TWS reductions significantly exacerbate both the frequency and the magnitude of wildfires. This relationship persists after controlling for variables such as the weather controls in the Fire Weather Index, seasonality, vegetation types, and common temporal trends, such as advancements in wildfire management. These findings remain robust after separately controlling for temperature, precipitation, SPEI, and varying definitions of wildfire occurrence based on burned areas.[11]

Specifically, a 1-standard-deviation increase in the rate of TWS loss leads to a 27 percent higher chance of wildfire occurrence and a 46 percent increase in the size of burned areas. There is large variation in the response function across climate zones and biomes that is likely driven by differences in land use and forest cover (Jones et al. 2022).[12] Notably, the effect is more than twice as pronounced in areas with temperate, winter-dry climates and in equatorial, summer-dry regions; conversely, the effect is negligible in arid climates.

The impact of TWS change on the occurrence of wildfires extends beyond the immediate period. The effect of TWS change is most pronounced within the month of fire activity. This effect diminishes when considering TWS changes from preceding months; however, the impact of TWS remains statistically significant for at least five months before fire activity. This finding indicates that reduced TWS not only increases the likelihood of fires in the current month but also affects future fire incidents.

In addition to physical factors such as land and vegetation species, socioeconomic development, wildfire prevention measures, and community practices play a critical role in determining the effect of freshwater availability on wildfires (World Bank 2023).

Using grid-level measures of GDP per capita and the Human Development Index (HDI) as indicators of socioeconomic development, locations are categorized into four quartiles on the basis of their GDP per capita or HDI measures in 2000. Results indicate a significant heterogeneous effect of TWS on wildfires. The most substantial adverse impact of TWS decline is observed in areas in which GDP per capita and HDI fall within the lowest quartiles. Notably, the effect of TWS changes in the lowest quartile is six to nine times greater than that observed in the next quartile.

This heterogenous effect is not driven by differences in land cover and climate zones that might correlate with income levels. A similar pattern is seen in each of the 29 climate zones and 14 biomes for which the GDP and HDI quartiles are defined.

These findings indicate that low-income regions are particularly vulnerable to the adverse effect of freshwater depletion on wildfires. In addition, communities in low-income regions and countries often have limited resources and capacity to recover from wildfires. Consequently, the combination of high vulnerability and limited coping capacity will lead to a disproportionately higher negative impact of TWS depletion on wildfires in low-income communities. However, the strong mediation effect of income and human development suggests an important role for policy interventions in reducing the negative impact of water scarcity and the associated risk of wildfire occurrence. Such interventions may include investing in fire suppression measures and management of wildland-urban interfaces.

Water and biodiversity

Biodiversity is vital for maintaining the health and balance of ecosystems on Earth. A rich variety of species ensures that ecosystems can adapt to changes, recover from disturbances, and continue providing essential services such as pollination, water purification, and carbon storage (Cardinale et al. 2012). Biodiversity also provides cultural, aesthetic, and economic value, offering resources for medicine, agriculture, and innovation (Newman and Cragg 2020). Conversely, the loss of biodiversity has significant consequences for the planet, including disruptions to food chains, changes in nutrient cycling, and decreases in resilience to environmental disturbances.

Biodiversity and water availability are linked through a reinforcing feedback loop. Healthy ecosystems, particularly wetlands, forests, and grasslands, play a key role in regulating local and regional water cycles by enhancing soil moisture retention, improving groundwater recharge, and influencing precipitation patterns (Brauman et al. 2007; IPBES 2019). In turn, reduced water availability can degrade ecosystems, driving biodiversity loss, which further impairs the ecosystem's capacity to regulate water, perpetuating a vicious cycle of ecological decline (Acreman et al. 2013). This feedback is particularly concerning in biodiversity hot spots, where ecosystems are already under severe pressure.

Biodiversity hot spots are regions that combine exceptionally high species richness with particularly severe threats to their ecosystem, making them global conservation priorities. Specifically, a biodiversity hot spot is characterized by the presence of at least 1,500 endemic species of vascular plants and a loss of at least 70 percent of its original native vegetation

(Hoffman et al. 2016; Marchese 2015; Myers et al. 2000). Currently, there are 36 globally recognized biodiversity hot spots. Although they cover only approximately 2.5 percent of Earth's surface, they support about 60 percent of the world's plant, bird, mammal, reptile, and amphibian species.[13]

An analysis overlaying the spatial distribution of biodiversity hot spots with observed TWS trends reveals that at least 17 of the 36 globally recognized hot spots are experiencing sustained reductions in freshwater availability (refer to map 2.3). Many hot spots across Central Asia, South America, and Southern Europe exhibit severe TWS loss, shown in deep red areas, signaling chronic drying. Among the most affected regions is the Cerrado, which stands out with an alarming loss of −17.1 km³ per year, followed by the Himalayas (−13.6 km³ per year), the Mediterranean Basin (−9.6 km³ per year), the Caucasus (−9.3 km³ per year), and Indo-Burma (−5.5 km³ per year). These findings highlight that some of the planet's most biologically rich ecosystems are also experiencing the most acute and sustained freshwater depletion. In these regions, even slight shifts in the water balance can disrupt species composition, ecosystem productivity, and critical ecological processes.

Almost half of global biodiversity hot spots are experiencing persistent drying.

MAP 2.3 Trends in TWS across global biodiversity hot spots

Source: World Bank.
Note: The map plots the annual rate of TWS change in biodiversity hot spots. Red denotes decreasing TWS; blue denotes increasing TWS. Notably, some hot spots in Southeast Asia and Sub-Saharan Africa display increasing TWS. However, this trend may indicate intensified flooding or human interventions, such as the development of large-scale dams, which can also devastate biodiversity. TWS = terrestrial water storage

In contrast, some hot spots display positive TWS trends, shown in blue on map 2.3, suggesting increased water availability. Notable examples include Eastern Afromontane (+11.0 km^3 per year), Sundaland (+9.3 km^3 per year), and the Guinean forests of West Africa (+2.3 km^3 per year). However, such increases may indicate intensified wet season flooding or human interventions, such as large-scale water infrastructure development, rather than natural ecosystem stability. Both chronic water loss and abrupt water surpluses could destabilize ecosystems, affecting species composition, habitat connectivity, and long-term ecological functioning (IPBES 2019).[14]

Although the correlation between TWS loss and biodiversity loss does not establish causality, the co-occurrence of water stress and ecological decline in these hot spots points to a dangerous reinforcing cycle. Water scarcity is recognized as a key driver of declining biological productivity and species richness (Wrubel and Parker 2018). Studies have identified multiple pathways through which reduced water availability influences species distribution and abundance.

First, water scarcity affects freshwater ecosystems and threatens water-sensitive habitats such as montane forests, wetlands, and tropical cloud forests (Shrestha and Devkota 2010). For example, in the Mediterranean Basin hot spot, declining rainfall and increased evapotranspiration have led to measurable decreases in forest biomass, threatening endemic oak species and reducing habitat for emblematic species (Peñuelas and Sardans 2021). Similarly, in the Mountains of Central Asia hot spot, glaciers that supply critical dry-season flows to montane river systems are retreating, and general dryness is threatening aquatic species and riparian vegetation (Seim et al. 2016; Xu, Wang, and Zhang 2016). In the Eastern Himalayan hot spot, shrinking water availability has already contributed to the decline of aquatic species, such as Himalayan river fish and amphibians adapted to high-elevation streams (Chettri et al. 2010).

Second, increasing water scarcity intensifies competition between human water demands and conservation efforts in biodiversity hot spots. In the Irano-Anatolian hot spot, approximately 1,500 km^2 of ecologically valuable swamps have been drained to support irrigation, water extraction, and agricultural expansion (Şekercioğlu et al. 2011). In the Cerrado hot spot of Brazil (refer to box 3.2), rapid expansion of irrigated agriculture and farming places severe pressure on local aquifers and surface water, fragmenting and patching native ecosystems and thus pushing species into shrinking habitat (Strassburg et al. 2017).[15] Similar tensions exist in the Atlantic Forest hot spot, where forest remnants, which protect about

60 percent of Brazil's terrestrial biodiversity (CEPF 2001), supply water to nearly half of Brazil's population. Declining water availability, in turn, forces difficult trade-offs between ecosystem protection and human water uses (Ribeiro et al. 2011; Tabarelli et al. 2005).

Finally, as discussed in the previous section, drying trends increase wildfire risk through vegetation flammability and reduced moisture in organic soils, particularly in fire-prone ecosystems such as the Mediterranean Basin, the California Floristic Province, and the Cerrado. For example, uncontrolled wildfires during dry spells have already led to substantial ecosystem losses in the Mesoamerica hot spot, with fire regimes increasingly altered by reduced water availability and vegetation stress (del Castillo and Rivera-García 2022).

Global biodiversity hot spots are particularly sensitive to drying-induced wildfires. Analysis for this report shows that a 1-standard-deviation decrease in TWS raises the probability of wildfire occurrence in biodiversity hot spots by 50 percent, compared with just 11 percent outside these areas.

Continental drying also affects freshwater ecosystems and ecological interactions. Wetlands, seasonal rivers, and groundwater-fed habitats, which serve as critical refuges for biodiversity, are especially vulnerable to continental drying. In the Cerrado, drying trends have contributed to the decline of vernal pools essential for amphibian reproduction (Colli, Vieira, and Dianese 2020); in the Mekong Basin, falling water levels from drought and human withdrawals have disrupted fish spawning cycles and altered food web structures (Ziv et al. 2012). Beyond freshwater ecosystems, drying trends also disrupt plant-pollinator networks and trophic interactions, which in turn affect the structure and services of ecosystems. As water availability declines, many plant species experience reduced flowering, lower nectar production, and altered phenological cycles, with cascading effects on pollinator and biological communities (Harrison 2000). At the same time, invasive drought-adapted species, such as grasses and shrubs, are increasingly expanding into drying regions, displacing native flora and altering fire regimes (Bradstock 2010).

Moreover, continental drying might also have profound repercussions on the Earth system. Most notable is the threat of tropical rainforest dieback because of prolonged droughts and increasing temperatures; this process might activate planetary tipping points. The Amazon is considered a potential tipping element of the Earth system; crossing a threshold of forest loss could trigger large-scale, irreversible shifts with repercussions in the global climate and biophysical system. However, in this region, droughts in 2005, 2010, 2015, and 2023–24 have already caused widespread tree mortality, shifting the Amazon's role from a net carbon sink to a carbon source in certain years while also affecting the water cycle (Gatti et al. 2021; Phillips et al. 2009). At the same time, Amazon dieback is associated with weakened tree resilience and altered species composition, favoring lower-biomass vegetation rather than water-dependent canopy trees (Brienen et al. 2015). Beyond activating critical regional thresholds, declining water levels may trigger cascading failures, whereby ecosystemic degradations accelerate hydrological instability, further deepening water scarcity, and destabilize the functioning of key global systems.

Conclusion

This chapter analyzes the widespread impacts of continental drying on people, prosperity, and the planet. In addition, it identifies patterns of resilience and vulnerability, and offers insights into potential adaptation mechanisms—whether through infrastructure development, economic diversification, or strategic trade relationship. The next chapter explores emerging trends of water demand to further pinpoint vulnerability hot spots—regions facing the dual crisis of increasing water demand and declining supply—and potential adaptation measures through improved demand-side management.

Notes

1. Various thresholds have been used to assess the severity of dry events, specifically declines of 1.0, 1.5, and 2.0 standard deviations below long-term average water availability. The results remain consistent across all thresholds; however, higher thresholds reveal stronger impacts of dry shocks on employment outcomes. In this report, results are presented using the 1.5-standard-deviation threshold as the basis for analysis.
2. An estimated 9 million jobs are created annually in Sub-Saharan Africa (Abdychev et al. 2018; IMF 2024).
3. Dry and hot regions are defined as areas with annual precipitation below 75 mm and an annual average temperature above 30°C. Moderately dry and moderately hot regions have annual precipitation between 75 mm and 150 mm and average temperatures ranging from 10°C to 20°C. Mildly dry and wet regions are those with annual precipitation exceeding 150 mm.
4. Berrittella et al. (2007) show that, when water is misallocated across sectors and regions, water supply constraints could lead to welfare gain by improving allocative efficiency. Furthermore, this welfare gain may more than offset the welfare losses due to the resource constraint.
5. Technical appendixes A through E are available online at https://hdl.handle.net/10986/43683.

6. The difference between the impact of SPEI and precipitation may be due to the fact that 1 standard deviation in SPEI is associated with a 155 mm variation in precipitation; 1 standard deviation in precipitation was about 95 mm from 1951 to 2022. It may also indicate that, given a similar level of decline, a decrease in soil moisture proxied by SPEI has a larger detrimental impact on crop growth than a similar-scale decline in precipitation.
7. This effect also diminishes in moderately hot regions and becomes negative in extremely cold regions, where precipitation often falls as snow, thus inhibiting vegetation growth.
8. Warmer temperatures are detrimental for plant growth in areas where the annual average temperature exceeds 5°C, but they are beneficial in colder regions.
9. Notably, the finding reflects the potential global cost of water scarcity in India, rather than a direct measurement of current annual losses.
10. California Department of Forestry and Fire Protection, Statistics for the 2020 fire season, https://www.fire.ca.gov.
11. In the baseline model, a grid cell with a burned area exceeding 21.44 hectares (ha), the smallest recorded in the data set, is counted as a wildfire occurrence. For a robustness check, stricter criteria are applied by restricting the wildfire occurrence sample to include areas with burned regions surpassing 100 ha, 1,000 ha, or 10,000 ha, respectively.
12. Climate zones segment the world on the basis of primary climate characteristics (equatorial, arid, temperate, snow, and polar) and their associated temperature and precipitation patterns. In this analysis, climate zones as delineated in the Köppen-Geiger climate classification (1976–2000) are based on Kottek et al. (2006). Biome definitions are from the ecoregion classification of Dinerstein et al. (2017). This categorization is rooted in the unique biological diversity each region represents, and it is further subdivided into 14 types of biomes (seven of which are forested) in which plant communities exhibit structural similarities (although species sets may differ). The map and the definitions of ecoregions and biomes can be viewed at https://ecoregions.appspot.com/.
13. Biodiversity-rich areas extend beyond the formal boundaries of these hot spots, with significant ecological value found in regions such as Sub-Saharan Africa that are not always captured by hot spot classification.
14. Many other pressures, such as land use change, habitat fragmentation, or direct exploitation, also threaten biodiversity. Furthermore, although TWS offers a valuable integrated signal of water availability, its relevance for biodiversity depends on the specific water source in question, such as surface water, soil moisture, or groundwater, each of which has distinct ecological implications. Future work should integrate hydrological disaggregation, high-resolution ecological data, and spatially explicit approaches to model and better capture local dynamics. Expanding the scope of analysis beyond formal hot spot boundaries and incorporating functional diversity constitute another key area for future research.
15. *Patching* refers to the creation of small, disconnected patches of habitat that are remnants of the original ecosystem. Smaller patches may not support viable populations of certain species. For example, patches are more susceptible to external pressures, such as invasive species or climate change. In addition, the interaction between species (for example, predator-prey relationships) may change, disrupting the balance of the ecosystem.

References

Abdychev, A. , C. Alonso, E. Alper, D. Desruelle, S. Kothari, Y. Liu, M. Perinet, S. Rehman, A. Schimmelpfennig, and P. Sharma. 2018. "The Future of Work in Sub-Saharan Africa." Departmental Paper 2018/018, International Monetary Fund, Washington, DC.

Acreman, M., J. Fisher, C. Stratford, D. J. Mould, and J. O. Mountford. 2013. "Wetlands and Water Framework Directive: Mechanisms and Tools for Achieving Good Ecological Status." *Hydrology and Earth Systems Science 17* (6): 2615–32.

Adger, W. Neil, Suraje Dessai, Marisa Goulden, Mike Hulme, Irene Lorenzoni, Donald R. Nelson, Lars Otto Naess, Johanna Wolf, and Anita Wreford. 2009. "Are There Social Limits to Adaptation to Climate Change?" *Climatic Change* 93: 335–54.X

Agnolucci, P., C. Rapti, P. Alexander, V. De Lipsis, R. A. Holland, F. Eigenbrod, and P. Ekins. 2020. "Impacts of Rising Temperatures and Farm Management Practices on Global Yields of 18 Crops." *Nature Food* 1 (9): 562–71.

Agrawal, A. 2008. "The Role of Local Institutions in Adaptation to Climate Change." Paper prepared for the Social Dimensions of Climate Change, Social Development Department, World Bank, Washington DC, March 5-6, 2008. https://open knowledge.worldbank.org/server/api/core/bitstreams/69ccd6cd-9b4f-5292-ae50 -a04d93f7ce2a/content.

Allan, J. A. 1998. "Virtual Water: A Strategic Resource Global Solutions to Regional Deficits." *Ground Water* 36 (4): 545–6. https://doi.org/10.1111/j.1745-6584.1998. tb02825.x.

Alsan, M., and C. Goldin. 2019. "Watersheds in Child Mortality: The Role of Effective Water and Sewerage Infrastructure, 1880–1920." *Journal of Political Economy* 127 (2): 586–638.

Asher, S., T. Lunt, R. Matsuura, and P. Novosad. 2021. "Development Research at High Geographic Resolution: An Analysis of Night-Lights, Firms, and Poverty in India Using the Shrug Open Data Platform." *World Bank Economic Review* 35 (4): 845–71.

Beillard, M. J., and S. K. Singh. 2022. "India: Extreme Temperature Scorch Indian Wheat Production." Report No. IN2022-0045, U.S. Department of Agriculture, Foreign Agricultural Service, Washington, DC.

Below, T. B., K. D. Mutabazi, D. Kirschke, C. Francke, S. Sieber, R. Sievert, and K. Tscherning. 2012. "Can Farmers' Adaptation to Climate Change Be Explained by Socio-Economic Household-Level Variables?" *Global Environmental Change 22* (1): 223–5.

Berrittella, M., A. Y. Hoekstra, K. Rehdanz, R. Roson, and R. S. J. Tol. 2007. "The Economic Impact of Restricted Water Supply: A Computable General Equilibrium Analysis." *Water Research* 41 (8): 1799–813.

Blakeslee, D., R. Fishman, and V. Srinivasan. 2020. "Way Down in the Hole: Adaptation to Long-Term Water Loss in Rural India." *American Economic Review* 110 (1): 200–24.

Boccaletti, Giulio. 2021. *Water: A Biography*. New York: Pantheon Books.

Boudot-Reddy, C., and A. Butler. 2025. "Watering the Seeds of the Rural Economy: Evidence From Groundwater Irrigation in India." *World Bank Economic Review* 39 (3): 571–91. https://doi.org/10.1093/wber/lhae041.

Bradstock, R. A. 2010. "A Biogeographic Model of Fire Regimes in Australia: Current and Future Implications." *Global Ecology and Biogeography* 19 (2): 145–58.

Brauman, K. A., G. C. Daily, T. K. E. Duarte, and H. A. Mooney. 2007. "The Nature and Value of Ecosystem Services: An Overview Highlighting Hydrologic Services." *Annual Review of Environment and Resources* 32 (1): 67–98.

Braun, A. C., F. Faßnacht, D. Valencia, and M. Sepulveda. 2021. "Consequences of Land-Use Change and the Wildfire Disaster of 2017 for the Central Chilean Biodiversity Hot spot." *Regional Environmental Change* 21 (2): 37.

Brienen, R. J. W., O. L. Phillips, T. R. Feldpausch, E. Gloor, T. R. Baker, J. Lloyd, G. Lopez-Gonzalez, et al. 2015. "Long-Term Decline of the Amazon Carbon Sink." *Nature* 519 (7543): 344–48.

Briscoe, J., and R. Malik. 2006. *India's Water Economy: Bracing for a Turbulent Future*. New Delhi: Oxford University Press.

Brown, C., R. Meeks, Y. Ghile, and K. Hunu. 2013. "Is Water Security Necessary? An Empirical Analysis of the Effects of Climate Hazards on National-Level Economic Growth." *Philosophical Transactions of the Royal Society A: Mathematical, Physical and Engineering Sciences 371* (2002): 20120416. https://doi.org/10.1098/rsta.2012.0416.

Burke, M., J. M. Tront, K. N. Lyon, W. Rex, M. I. Castera Errea, M. C. Varughese, J. T. Newton, et al. 2023. *What the Future Has in Store: A New Paradigm for Water Storage—Overview for Policy Makers*. Washington, DC: World Bank. http://documents.worldbank.org/curated/en/099203002012336127.

Cardinale, B. J., J. E. Duffy, A. Gonzalez, D. U. Hooper, C. Perrings, P. Venail, A. Narwani, et al. 2012. "Biodiversity Loss and Its Impact on Humanity." *Nature* 486 (7401): 59–67. https://doi.org/10.1038/nature11148.

Carleton, T. A., and S. M. Hsiang. 2016. "Social and Economic Impacts of Climate." *Science* 353 (6304): aad9837. https://doi.org/10.1126/science.aad9837.

Carleton, T., A. Jina, M. Delgado, M. Greenstone, T. Houser, S. Hsiang, A. Hultgren, et al. 2022. "Valuing the Global Mortality Consequences of Climate Change Accounting for Adaptation Costs and Benefits." *Quarterly Journal of Economics* 137 (4): 2037–105.

CEPF (Critical Ecosystem Partnership Fund). 2001. "Ecosystem Profile: Atlantic Forest Biodiversity Hotspot: Brazil." CEPF. https://www.cepf.net/resources/ecosystem-profile-documents/atlantic-forest-ecosystem-profile-2001.

Chandanpurkar, Hrishikesh A., James S. Famiglietti, Kaushik Gopalan, David N. Wiese, Yoshihide Wada, Kaoru Kakinuma, John T. Reager, and Fan Zhang. 2025. "Unprecedented Continental Drying, Shrinking Freshwater Availability, and Increasing Land Contributions to Sea Level Rise." Science Advances, 11: eadx0298. https://doi.org/10.1126/sciadv.adx0298. Background paper for this report.

Chettri, N., E. Sharma, B. Shakya, R. Thapa, B. Bajracharya, K. Uddin, K. P. Oli, and D. Choudhury. 2010. *Biodiversity in the Eastern Himalayas: Status, Trends and Vulnerability to Climate Change*. Kathmandu: International Centre for Integrated Mountain Development.

Clement, Viviane Wei Chen, Kanta K. Rigaud, Alex de Sherbinin, Bryan R. Jones, Susana Adamo, Jacob Schewe, Nian Sadiq, and Elham Shirin Shabahat. 2021. *Groundswell Part II: Acting on Internal Climate Migration*. Washington, DC: World Bank Group. https://hdl.handle.net/10986/36248.

Colli, G. R., C. R. Vieira, and J. C. Dianese. 2020. "Biodiversity and Conservation of the Cerrado: Recent Advances and Old Challenges." *Biodiversity and Conservation* 29 (5): 1465–75.

Coop, J. D., S. A. Parks, C. S. Stevens-Rumann, S. D. Crausbay, P. E. Higuera, M. D. Hurteau, A. Tepley, et al. 2020. "Wildfire-Driven Forest Conversion in Western North American Landscapes." *BioScience* 70 (8): 659–73.

Core Writing Team, Hoesung Lee, and José Romero, eds. 2023. *Climate Change 2023: Synthesis Report*. Geneva: Intergovernmental Panel on Climate Change. https://doi.org/10.59327/IPCC/AR6-9789291691647.

Dalgaard, C. J., A. S. B. Knudsen, and P. Selaya. 2020. "The Bounty of the Sea and Long-Run Development." *Journal of Economic Growth* 25 (3): 259–95.

Damania, Richard, Sébastien Desbureaux, Aude-Sophie Rodella, Jason Russ, and Esha Zaveri. 2017. *Uncharted Waters: The New Economics of Water Scarcity and Variability*. Washington, DC: World Bank.

Damania, R., S. Desbureaux, and E. Zaveri. 2020. "Does Rainfall Matter for Economic Growth? Evidence From Global Sub-National Data (1990–2014)." *Journal of Environmental Economics and Management* 102: 102335.

Das, M. B., C. F. Fisiy, and R. Kyte. 2013. *Inclusion Matters: The Foundation for Shared Prosperity*. New Frontiers of Social Policy. Washington, DC: World Bank.

del Castillo, R. F., and R. Rivera-García. 2022. "Biophysical and Biocultural Upheavals in Mesoamerica, a Conservation Perspective: Mountains, Maize-Milpa, and Globalization." *Frontiers in Forests and Global Change* 5: 763009.

Dell, M., B. F. Jones, and B. A. Olken. 2012. "Temperature Shocks and Economic Growth: Evidence from the Last Half Century." *American Economic Journal: Macroeconomics* 4 (3): 66–95.

Deschênes, O., and M. Greenstone. 2011. "Climate Change, Mortality, and Adaptation: Evidence from Annual Fluctuations in Weather in the US." *American Economic Journal: Applied Economics* 3 (4): 152–85.

Diamond, L. J. 1997. "Civil Society and the Development of Democracy." Working Paper, Center for Advanced Study in the Social Sciences, Madrid.

Dinerstein, E., D. Olson, A. Joshi, C. Vynne, N. D. Burgess, E. Wikramanayake, N. Hahn, et al. 2017. "An Ecoregion-Based Approach to Protecting Half the Terrestrial Realm." *BioScience* 67 (6): 534–45.

Doerr, S. H., and C. Santín. 2016. "Global Trends in Wildfire and Its Impacts: Perceptions versus Realities in a Changing World." *Philosophical Transactions of the Royal Society B: Biological Sciences* 371 (1696): 20150345.

Eaton, J., and S. Kortum. 2002. "Technology, Geography, and Trade." *Econometrica* 70 (5): 1741–79.

Ellis, T. M., D. M. Bowman, P. Jain, M. D. Flannigan, and G. J. Williamson. 2022. "Global Increase in Wildfire Risk Due to Climate-Driven Declines in Fuel Moisture." *Global Change Biology* 28 (4): 1544–59.

Emran, M. G. I., R. Barma, A. H. Khan, and M. Roy. 2024. "Reasons behind the Water Crisis and Its Potential Health Outcomes." *European Journal of Development Studies* 4 (6): 16–24. https://doi.org/10.24018/ejdevelop.2024.4.6.389.

FAO (Food and Agriculture Organization of the United Nations). 2024. *The State of Agricultural Commodity Markets 2024*. Rome: FAO.

Fishman, R. 2018. "Groundwater Depletion Limits the Scope for Adaptation to Increased Rainfall Variability in India." *Climatic Change 147*: 195–209.

Friedl, M., and D. Sulla-Menashe. 2022. "MODIS/Terra+Aqua Land Cover Type Yearly L3 Global 500m SIN Grid V061." Data set. Sioux Falls, SD: NASA EOSDIS Land Processes Distributed Active Archive Center.

FSIN (Food Security Information Network). 2018. *Global Report on Food Crises 2018*. FSIN. https://www.fao.org/fileadmin/user_upload/fsin/docs/global_report/2018/GRFC_2018_Full_report_EN.pdf.

Galor, O., and Ö. Özak. 2016. "The Agricultural Origins of Time Preference." *American Economic Review* 106 (10): 3064–103.

Gatti, L. V., J. Melack, L. S. Basso, N. Restrepo-Coupe, S. Pangala, S. R. Saleska, L. Aragão, O. L. Phillips, and D. Armenteras. 2021. "Cross-Chapter 1: The Amazon Carbon Budget." In *Amazon Assessment Report—Part 1: The Amazon as a Regional Entity of the Earth System*, edited by Carlos Nobre, Ana Encalada, Elizabeth Anderson, Fabiola Heredia Roca Alcazar, Mercedes Bustamante, Carolina Mena, Marielos Peña-Claros, et al. New York: United Nations Sustainable Development Solutions Network.

GCEW (Global Commission on the Economics of Water). 2024. *The Economics of Water: Valuing the Hydrological Cycle as a Global Public Good*. Colombo, Sri Lanka: GCEW, International Water Management Institute.

Giglio, L., W. Schroeder, J. V. Hall, and C. O. Justice. 2021. *MODIS Collection 6 and Collection 6.1 Active Fire Product User's Guide*. Washington, DC: National Aeronautical and Space Administration.

Giuliano, P., and N. Nunn. 2021. "Understanding Cultural Persistence and Change." *Review of Economic Studies* 88 (4): 1541–81.

Haberl, H., K. H. Erb, F. Krausmann, V. Gaube, A. Bondeau, C. Plutzar, S. Gingrich, W. Lucht, and M. Fischer-Kowalski. 2007. "Quantifying and Mapping the Human Appropriation of Net Primary Production in Earth's Terrestrial Ecosystems." *Proceedings of the National Academy of Sciences* 104 (31): 12942–7.

Harari, M., and E. L. Ferrara. 2018. "Conflict, Climate, and Cells: A Disaggregated Analysis." *Review of Economics and Statistics* 100 (4): 594–608.

Harrison, N. E. 2000. *Constructing Sustainable Development*. New York: SUNY Press.

Henderson, A. D., A. C. Asselin-Balençon, M. Heller, L. Lessard, S. Vionnet, and O. Jolliet. 2017. "Spatial Variability and Uncertainty of Water Use Impacts From US Feed and Milk Production." *Environmental Science & Technology* 51 (4): 2382–91.

Hibbs, D. A., Jr., and O. Olsson. 2004. "Geography, Biogeography, and Why Some Countries Are Rich and Others Are Poor." *Proceedings of the National Academy of Sciences* 101 (10): 3715–20.

Hoffman, O., N. de Falco, H. Yizhaq, and B. Boeken. 2016. "Annual Plant Diversity Decreases across Scales Following Widespread Ecosystem Engineer Shrub Mortality." *Journal of Vegetation Science* 27 (3): 578–86.

Hornbeck, R. 2012. "The Enduring Impact of the American Dust Bowl: Short- and Long-Run Adjustments to Environmental Catastrophe." *American Economic Review* 102 (4): 1477–507.

Hornbeck, R., and P. Keskin. 2014. "The Historically Evolving Impact of the Ogallala Aquifer: Agricultural Adaptation to Groundwater and Drought." *American Economic Journal: Applied Economics* 6 (1): 190–219.

Hsiang, S. 2016. "Climate Econometrics." *Annual Review of Resource Economics* 8 (1): 43–75.

Hunt, E. D., F. Femia, C. Werrell, J. Christian, J. A. Otkin, J. Basara, M. Anderson, T. R. White, C. Hain, R. Randall, and K. McGaughey. 2021. "Agricultural and Food Security Impacts from the 2010 Russia Flash Drought." *Weather and Climate Extremes* 34: 100383. https://doi.org/10.1016/j.wace.2021.100383.

Hunter, P. R., A. M. MacDonald, and R. C. Carter. 2010. "Water Supply and Health." *PLoS Medicine* 7 (11): e1000361.

Hyland, M., and J. Russ. 2019. "Water as Destiny—The Long-Term Impacts of Drought in Sub-Saharan Africa." *World Development* 115: 30–45.

IMF (International Monetary Fund). 2024. *The Clock Is Ticking: Meeting Sub-Saharan Africa's Urgent Job Creation Challenge.* Regional Economic Outlook Notes: Sub-Saharan Africa. Washington, DC: IMF. https://www.imf.org/-/media/Files/Publications/REO/AFR/2024/October/English/JobsNote.ashx.

Immerzeel, W. W., A. F. Lutz, M. Andrade, A. Bahl, H. Biemans, T. Bolch, S. Hyde, et al. 2020. "Importance and Vulnerability of the World's Water Towers." *Nature 577:* 364–9.

Immerzeel, W. W., L. P. Van Beek, and M. F. Bierkens. 2010. "Climate Change Will Affect the Asian Water Towers." *Science* 328 (5984): 1382–5.

IPBES (Intergovernmental Science-Policy Platform on Biodiversity and Ecosystem Services). 2019. *Summary for Policymakers of the Global Assessment Report on Biodiversity and Ecosystem Services of the Intergovernmental Science-Policy Platform on Biodiversity and Ecosystem Services.* Bonn: IPBES Secretariat.

Jaafar, H. H., and F. A. Ahmad. 2015. "Crop Yield Prediction from Remotely Sensed Vegetation Indices and Primary Productivity in Arid and Semi-Arid Lands." *International Journal of Remote Sensing* 36 (18): 4570–89.

Jolly, W. M., M. A. Cochrane, P. H. Freeborn, Z. A. Holden, T. J. Brown, G. J. Williamson, and D. M. Bowman. 2015. "Climate-Induced Variations in Global Wildfire Danger from 1979 to 2013." *Nature Communications* 6 (1): 7537.

Jones, M. W., J. T. Abatzoglou, S. Veraverbeke, N. Andela, G. Lasslop, M. Forkel, A. J. P. Smith, et al. 2022. "Global and Regional Trends and Drivers of Fire under Climate Change." *Reviews of Geophysics* 60 (3): e2020RG000726.

Khan, Amjad, Landry Kuate, Roland Pongou, and Fan Zhang. 2024. "Weather, Water, and Work: Climatic Water Variability and Labor Market Outcomes in Sub-Saharan Africa." Background paper for this report and Policy Research Working Paper 10823, World Bank, Washington, DC.

Kiely, L., D. V. Spracklen, S. R. Arnold, E. Papargyropoulou, L. Conibear, C. Wiedinmyer, C. Knote, and H. A. Adrianto. 2021. "Assessing Costs of Indonesian Fires and the Benefits of Restoring Peatland." *Nature Communications* 12 (1): 7044.

Koolwal, G., and D. Van de Walle. 2013. "Access to Water, Women's Work, and Child Outcomes." *Economic Development and Cultural Change* 61 (2): 369–405.

Kottek, M., J. Grieser, C. Beck, B. Rudolf, and F. Rubel. 2006. "World Map of the Köppen-Geiger Climate Classification Updated." *Meterologische Zeitschrift* 15 (3): 259–63.

Kotz, M., A. Levermann, and L. Wenz. 2022. "The Effect of Rainfall Changes on Economic Production." *Nature* 601 (7892): 223–27.

Kubik, Z., and M. Maurel. 2023. "Water Scarcity and Migration: Evidence from the Middle East and North Africa." *Journal of Environmental Economics and Management* 118: 102785. https://doi.org/10.1016/j.jeem.2022.102785.

Law, B. E., J. T. Abatzoglou, C. R. Schwalm, D. Byre, N. Fann, and J. J. Nassikas. 2025. "Anthropogenic Climate Change Contributes to Wildfire Particulate Matter and Related Mortality in the United States." *Communications Earth & Environment* 6: 336. https://doi.org/10.1038/s43247-025-02314-0.

Lesk, C., E. Coffel, J. Winter, D. Ray, J. Zscheischler, S. I., Seneviratne, and R. Horton. 2021. "Stronger Temperature–Moisture Couplings Exacerbate the Impact of Climate Warming on Global Crop Yields." *Nature Food* 2: 683–91. https://doi.org/10.1038/s43016-021-00341-6.

Lu, Yang, Ting Yang, Jianyu Fu, and Wenlong Song. 2025. "Utility of the Standardized Precipitation Evapotranspiration Index (SPEI) to Detect Agricultural Droughts over China." *Journal of Hydrology: Regional Studies* 58: 102190.

Maccini, S., and D. Yang. 2009. "Under the Weather: Health, Schooling, and Economic Consequences of Early-Life Rainfall." *American Economic Review* 99 (3): 1006–26.

Marchese, C. 2015. "Biodiversity Hotspots: A Shortcut for a More Complicated Concept." *Global Ecology and Conservation* 3: 297–309. https://doi.org/10.1016/j.gecco.2014.12.008.

McDonald, R. I., K. Weber, J. Padowski, M. Flörke, C. Schneider, P. A. Green, T. Gleeson, et al. 2014. "Water on an Urban Planet: Urbanization and the Reach of Urban Water Infrastructure." *Global Environmental Change* 27: 96–105.

Meier, S., E. Strobl, R. J. R. Elliott, and N. Kettridge. 2023. "Cross-Country Risk Quantification of Extreme Wildfires in Mediterranean Europe." *Risk Analysis* 43 (9): 1745–62. https://doi.org/10.1111/risa.14075.

Michalopoulos, S., L. Putterman, and D. N. Weil. 2019. "The Influence of Ancestral Lifeways on Individual Economic Outcomes in Sub-Saharan Africa." *Journal of the European Economic Association* 17 (4): 1186–231.

Monfreda, C., N. Ramankutty, and J. A. Foley. 2008. "Farming the Planet: 2. Geographic Distribution of Crop Areas, Yields, Physiological Types, and Net Primary Production in the Year 2000." *Global Biogeochemical Cycles* 22 (1): GB1022. https://doi.org/10.1029/2007GB002947.

Morton, J. F. 2007. "The Impact of Climate Change on Smallholder and Subsistence Agriculture." *Proceedings of the National Academy of Sciences* 104 (50): 19680–5.

Murdock, G. P. 1967. "Ethnographic Atlas: A Summary." *Ethnology* 6 (2): 109–236.

Musungu, Arnold, Zaneta Kubik, and Matin Qaim. 2023. "Drought Shocks and Labor Reallocation in Rural Africa: Evidence From Ethiopia." Discussion Papers on Economic Policy No. 334, Center for Development Research, Bonn.

Myers, N., R. A. Mittermeier, C. G. Mittermeier, G. A. Da Fonseca, and J. Kent. 2000. "Biodiversity Hot spots for Conservation Priorities." *Nature* 403 (6772): 853–58.

Nath, Isan. 2024. "Climate Change, the Food Problem, and the Challenge of Adaptation through Sectoral Reallocation." Working Paper, Federal Reserve Bank of San Francisco, San Francisco.

Newman, D. J., and G. M. Cragg. 2020. "Natural Products as Sources of New Drugs over the Nearly Four Decades from 1981 to 2019." *Journal of Natural Products* 83 (3): 770–803.

Nguyen, D. L., T. T. Nguyen, and U. Grote. 2023. "Shocks, Household Consumption, and Livelihood Diversification: A Comparative Evidence from Panel Data in Rural Thailand and Viet Nam." *Economic Change and Restructuring* 56 (5): 3223–55. https://doi.org/10.1007/s10644-022-09400-9.

Peñuelas, J., and J. Sardans. 2021. "Global Change and Forest Disturbances in the Mediterranean Basin: Breakpoints, Mechanisms, and Ecosystem Responses." *Science of the Total Environment* 781: 146738. https://doi.org/10.1016/j.scitotenv.2021.146738.

Phillips, O. L., L. E. Aragão, S. L. Lewis, J. B. Fisher, J. Lloyd, G. López-González, Y. Malhi, et al. 2009. "Drought Sensitivity of the Amazon Rainforest." *Science* 323 (5919): 1344–7.

Pingali, P. L. 2012. "Green Revolution: Impacts, Limits, and the Path Ahead." *Proceedings of the National Academy of Sciences* 109 (31): 12302–8. https://doi.org/10.1073/pnas.0912953109.

Pongou, R., M. Ezzati, and J. A. Salomon. 2006. "Household and Community Socioeconomic and Environmental Determinants of Child Nutritional Status in Cameroon." *BMC Public Health* 6: 1–19.

Prüss-Üstün, A., S. Bonjour, and C. Corvalán. 2008. "The Impact of the Environment on Health by Country: A Meta-Synthesis." *Environmental Health* 7: 7.

Putterman, L. 2008. "Agriculture, Diffusion, and Development: Ripple Effects of the Neolithic Revolution." *Economica* 75 (300): 729–48.

Ribeiro, M. C., A. C. Martensen, J. P. Metzger, M. Tabarelli, F. Scarano, and M. J. Fortin. 2011. "The Brazilian Atlantic Forest: A Shrinking Biodiversity Hot spot." In *Biodiversity Hot spots: Distribution and Protection of Conservation Priority Areas*, edited by F. E. Zachos and J. C. Habel, 405–34. Heidelberg, Germany: Springer

Rodella, A., E. Zaveri, and F. Bertone. 2023. *The Hidden Wealth of Nations: The Economics of Groundwater in Times of Climate Change*. Washington, DC: World Bank.

Rounce, D. R., R. Hock, F. Maussion, R. Hugonnet, W. Kochtitzky, M. Huss, E. Berthier, et al. 2023. "Global Glacier Change in the 21st Century: Every Increase in Temperature Matters." *Science* 379: 78–83.

Rudik, I., G. Lyn, W. Tan, and A. Ortiz-Bobea. 2022. "The Economic Effects of Climate Change in Dynamic Spatial Equilibrium." Paper presented at the GTAP Annual Conference on Global Economic Analysis, virtual, June 8–10.

Russ, J. 2020. "Water Runoff and Economic Activity: The Impact of Water Supply Shocks on Growth." *Journal of Environmental Economics and Management* 101: 102322.

Schlenker, W., W. Michael Hanemann, and A. C. Fisher. 2005. "Will US Agriculture Really Benefit from Global Warming? Accounting for Irrigation in the Hedonic Approach." *American Economic Review 95* (1): 395–406.

Şekercioğlu, Ç. H., S. Anderson, E. Akçay, R. Bilgin, Ö. E. Can, G. Semiz, C. Tavşanoğlu, et al. 2011. "Turkey's Globally Important Biodiversity in Crisis." *Biological Conservation* 144 (12): 2752–69.

Seim, A., G. Omurova, E. Azisov, K. Musuraliev, K. Aliev, T. Tulyaganov, and H. W. Linderholm. 2016. "Climate Change Increases Drought Stress of Juniper Trees in the Mountains of Central Asia." *PloS One* 11 (4): e0153888.

Shrestha, A. B., and L. P. Devkota. 2010. *Climate Change in the Eastern Himalayas: Observed Trends and Model Projections*. Kathmandu, Nepal: International Centre for Integrated Mountain Development.

Strassburg, B. B., T. Brooks, R. Feltran-Barbieri, A. Iribarrem, R. Crouzeilles, R. Loyola, A. E. Latawiec, et al. 2017. "Moment of Truth for the Cerrado Hot spot." *Nature Ecology and Evolution* 1 (4): 0099.

Tabarelli, M., L. P. Pinto, J. M. Silva, M. Hirota, and L. Bede. 2005. "Challenges and Opportunities for Biodiversity Conservation in the Brazilian Atlantic Forest." *Conservation Biology* 19 (3): 695–700.

USDA (U.S. Department of Agriculture). 2013. "Crop Production Down in 2012 due to Drought, USDA Reports." News Release, January 11, 2013. https://www.nass.usda.gov/Newsroom/archive/2013/01_11_2013.php.

Vicente-Serrano, Sergio M., Beguería, and Juan I. López-Moreno. 2010. "A Multiscalar Drought Index Sensitive to Global Warming: The Standardized Precipitation Evapotranspiration Index." *Journal of Climate* 23 (7): 1696–718.

Wang, Hongshuo, Jeffrey C. Rogers, and Darla K. Munroe. 2015. "Commonly Used Drought Indices as Indicators of Soil Moisture in China." *Journal of Hydrometeorology* 16 (3): 1397–408.

Wang, X., K. Studens, M. A. Parisien, S. W. Taylor, J. N. Candau, Y. Boulanger, and M. D. Flannigan. 2020. "Projected Changes in Fire Size from Daily Spread Potential in Canada over the 21st Century." *Environmental Research Letters* 15 (10): 104048.

Williams, A. P., J. T. Abatzoglou, A. Gershunov, J. Guzman-Morales, D. A. Bishop, J. K. Balch, and D. P. Lettenmaier. 2019. "Observed Impacts of Anthropogenic Climate Change on Wildfire in California." *Earth's Future* 7 (8): 892–910. https://doi.org/10.1029/2019EF001210.

World Bank. 2016. *"High and Dry: Climate Change, Water, and the Economy."* Video, May 2, 2016. World Bank. https://www.worldbank.org/en/topic/water/publication/high-and-dry-climate-change-water-and-the-economy.

World Bank. 2023. *Building Code Checklist for Fire Safety*. Washington, DC: World Bank. https://openknowledge.worldbank.org/handle/10986/40352.

Wrubel, E., and V. T. Parker. 2018. "Local Patterns of Diversity in California Northern Coastal Scrub." *Ecology and Evolution* 8 (15): 7250–60.

Xu, H. J., X. P. Wang, and X. X. Zhang. 2016. "Decreased Vegetation Growth in Response to Summer Drought in Central Asia From 2000 to 2012." *International Journal of Applied Earth Observation and Geoinformation* 52: 390–402.

Zaveri, Esha, Jason Russ, Amjad Khan, Richard Damania, Edoardo Borgomeo, and Anders Jägerskog. 2021. *Ebb and Flow. Vol. 1 of Water, Migration, and Development*. Washington, DC: World Bank. https://hdl.handle.net/10986/36089.

Zhang, F., and C. Borja-Vega. 2024. *Water for Shared Prosperity*. Washington, DC: World Bank. https://hdl.handle.net/10986/41575.

Ziv, G., E. Baran, S. Nam, I. Rodríguez-Iturbe, and S. A. Levin. 2012. "Trading-Off Fish Biodiversity, Food Security, and Hydropower in the Mekong River Basin." *Proceedings of the National Academy of Sciences* 109 (15): 5609–14.

CHAPTER 3

Water Vulnerability Hot Spots and Water Savings Potential

Key findings

- Global water consumption rose by 25 percent from 2000 to 2019, with about one-third of this increase occurring in regions experiencing drying.
- In drying regions, significant inefficiencies persist—about one-quarter of rainfed and one-third of irrigated crop water consumption are below global median efficiency levels.
- Improving the water use efficiency of inefficient producers of key crops to the global median level could lower irrigation water consumption by 137 billion cubic meters—equivalent to the annual water needs of 118 million people. The largest savings potential globally would be observed in South Asia and in the production of wheat and rice.
- Virtual water trade—the exchange of water embedded in traded agricultural and industrial commodities—has contributed to substantial annual global water savings, equivalent to 9.4 percent of global water consumption of 35 key crops. Eliminating distortions in agricultural input and output markets, boosting the efficiency of exporters, and integrating sustainable water management practices into trade agreements can further enhance the benefits of virtual water trade.

Introduction

The preceding chapters have highlighted the ongoing crisis of continental drying, marked by decreasing freshwater availability, and the significant economic and environmental impacts that may ensue. This chapter shifts

---

A reproducibility package is available for this book in the Reproducible Research Repository at https://reproducibility.worldbank.org/index.php/catalog/386.

the focus to the demand side of water. Although continental drying affects vast areas of global land, its impacts are shaped by local water consumption patterns. Regions with rapidly rising water consumption, low water use efficiency, and more water-intensive economic activities are more susceptible to the impact of continental drying. Agriculture is the largest water consumer globally. By combining water availability and agricultural water demand data, this chapter identifies water vulnerability hot spots and priority regions for policy interventions.

To analyze water consumption patterns, the analysis in this chapter uses terms and methodologies from the field of water footprint assessment (refer to box 3.1). The most recent comprehensive global water footprint assessment was conducted by Hoekstra and Mekonnen (2012), providing averages for 1996 to 2005. Given the significant shifts in water use since then, an updated water footprint estimate is needed.

Using a state-of-the-art crop growth model and the most up-to-date, globally comprehensive data on agriculture and on industrial and domestic water withdrawal, this report provides annual water footprint estimates for 169 countries from 2000 to 2019. The water footprint data are then overlaid with terrestrial water storage (TWS) trend data to identify the regions most susceptible to continental drying. Refer to Mialyk, Schyns et al. (2024) for the methodology, assumptions, and data used to estimate the water footprints.

BOX 3.1
Water footprint: Key terminology

A water footprint measures the appropriation of water, including soil moisture, surface water, and groundwater, to meet human needs. Although indicators such as water withdrawal or extraction measure gross water use, the water footprint accounts for return flows to the system. It therefore reflects net water use, also known as consumptive use.

A water footprint can be used to measure direct water consumption from a producer's perspective, such as water used to grow crops. It can also be used to assess indirect water consumption from a consumer's perspective, that is, water consumed in the form of food, energy, and industrial goods

(continued)

BOX 3.1
Water footprint: Key terminology *(continued)*

and services. In this context, the concept of a water footprint can be used to analyze the virtual water trade—the exchange of water embedded in traded agricultural and industrial commodities.

A water footprint has green, blue, and gray components. The green water footprint refers to water consumed from rainwater in the soil that does not become runoff or percolate to groundwater. It is the sole source of water in rain-fed production. The blue water footprint refers to water consumed from surface and groundwater bodies through evapotranspiration, product integration, or discharge into other river basins or the sea. The gray water footprint measures pollution by indicating the volume of freshwater required to dilute pollutants to meet water quality standards. Because of the lack of crop-specific water pollution data at the necessary scale and resolution, this analysis does not include the gray water footprint. Excluding degradative water use may result in an underestimation of the total water footprint.

The crop water footprint reported in this chapter reflects potential rather than actual values. Certain limitations in the simulation may lead to an under- or overestimation of water footprint. For instance, the simulation considers on-field crop water use only, disregarding other water-consuming activities upstream or downstream of the crop value chain, such as water losses during the transport of irrigation water from source to field and postharvest water requirements. Additionally, the simulated water footprint data estimate irrigation water needs on the basis of soil moisture rules. However, farmers in regions experiencing water scarcity may practice deficit irrigation, whereby crops receive less water than their full evapotranspiration needs. To address this issue, multiple sensitivity analyses were conducted to test modeling assumptions. Refer to Mialyk et al. (2024) for more details.

The use of remote sensing data offers a promising avenue to monitor the actual water footprints of crop production. Satellite-based sensors, such as those from the Sentinel-2 and Landsat programs, can track key indicators such as vegetation indexes, evapotranspiration rates, and soil moisture levels in near real time. Leveraging these technologies in future analyses can enhance the precision of crop water use assessments and improve water resource management.

Vulnerability hot spots: Rising demand and decreasing availability

At the local level, water footprint trends, shaped by water use efficiency and crop selection, vary significantly. These factors influence the region's water dependency and its future water resilience. This section combines TWS trends data and water footprints to identify the hot spot regions that are most vulnerable to continental drying. The vulnerability of a drying region is assessed on the basis of three factors: (a) whether the region is arid and undergoing a rapid increase in water footprint, (b) whether crop production in the drying region has low water use efficiency, and (c) whether the drying region is expanding the cultivation of more water-intensive crops.

Increasing water footprint

The analysis in this report indicates that the global water footprint of humanity grew by 25.3 percent between 2000 and 2019, rising from 7.2 trillion m^3 to 9.1 trillion m^3. Agriculture remains the dominant contributor to global human water consumption. In 2019, agriculture accounted for 98 percent of the global water footprint. In the agricultural sector, crop production accounted for 75 percent of the water footprint; livestock, including water used for fodder crops, grazing, drinking, and servicing, accounted for 23 percent of the total water footprint[1] (refer to figure 3.1).

Against the backdrop of continental drying, about one-third of the increase in the global water footprint has taken place in regions affected by drying. Regions experiencing rapid TWS decline and those showing a substantial increase in water consumption are in part of Argentina, southeast Brazil, Central America, northern China, a large swath of Eastern Europe, northern India, southeast Asia, and southwestern North America (refer to map 3.1). Most of these regions are experiencing worsening groundwater depletion.

These hot spot regions are predominantly cropland areas that are already water scarce. In 2019, 69 percent of global blue water consumption occurred in areas with moderate to severe water scarcity. Additionally, 94 percent of the increase in blue water consumption since 1990 has taken place in regions already experiencing water scarcity. Even historically water-abundant regions are facing growing threats of water scarcity because of rapid cropland expansion and unsustainable water management practices. The Cerrado region in Brazil is such a hot spot area (refer to box 3.2).

Between 2000 and 2019, the global water footprint of humanity grew by 25.3 percent. Agriculture is the largest water consumer.

FIGURE 3.1 Global water footprint, 2000–19

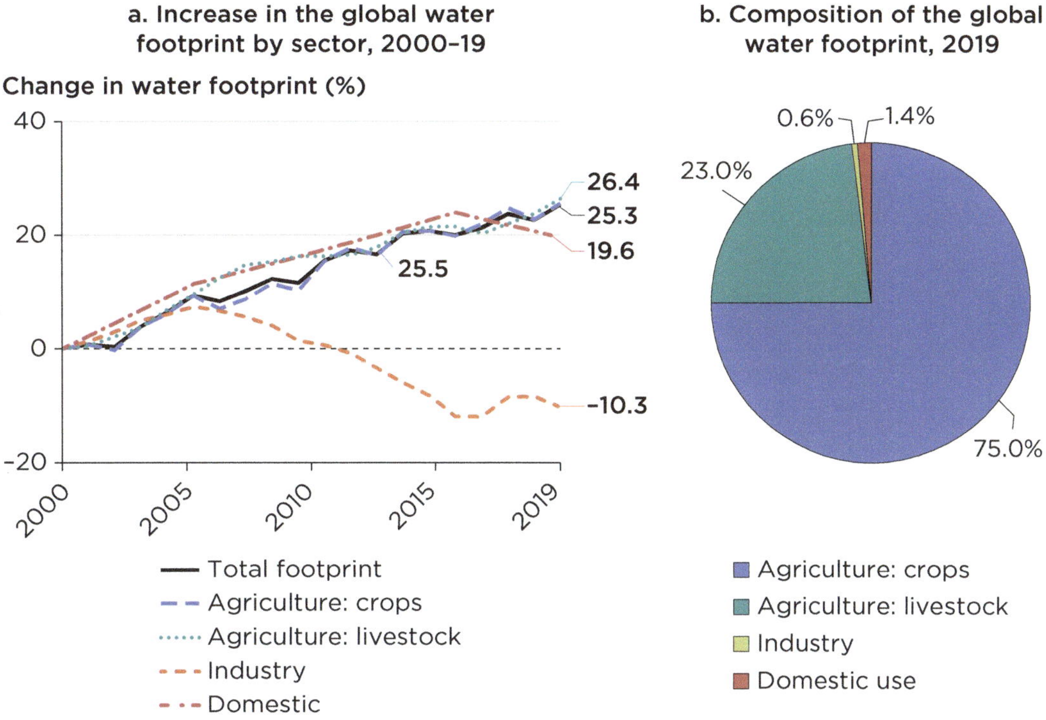

Source: World Bank.

Notes: In panel a, the lines denote in percentage points the change in water footprint relative to year 2000. The thick black line corresponds to the total water footprint whereas the dashed lines correspond to each sector. The numbers on the right denote the percentage increase in the water footprint in 2019 relative to 2000. In panel b, segments of the pie chart denote the proportion of the total water footprint that each sector represents.

BOX 3.2

Water crisis in the Cerrado

The Cerrado savanna is vast. Although less well-known than its neighbor, the Amazon, it covers 24 percent of Brazil, sprawling across the middle of the country. The Cerrado is South America's largest savanna and one of the most biodiverse savanna landscapes globally. It is home to 11,620 plant species and more than 200 animal species

(continued)

BOX 3.2
Water crisis in the Cerrado *(continued)*

(Spera et al. 2016). This ancient biodiversity is supported by groundwater and water recycling in the area's vegetation.

In recent decades, the Cerrado has been transformed into a center of Brazil's booming agribusiness. This agricultural expansion began in the 1960s, but the deforestation and associated land conversion have accelerated dramatically in recent decades. The Cerrado has lost half of its original cover of 2 million km^2 (Latrubesse et al. 2019). It is now Brazil's largest established agricultural area, generating 60 percent of its agricultural production.[a] Globally, it provides 12 percent of soybean production and 10 percent of beef exports (Rodrigues et al. 2022). As these industries grow, so too do human populations and the infrastructures needed to support them.

The removal of vegetation, coupled with the rapid development of irrigation-dependent agriculture, has put a huge strain on the Cerrado's water resources. Reduced vegetation raises ground temperatures and lowers humidity, causing lower rainfall. The depleted water supply has affected local wildlife as well as the human communities indigenous to the region. It also affects larger-scale water systems. The Urucuia Aquifer spans 120 km^2 of the Cerrado. Its levels are depleting primarily because of intensive water pumping (Rodrigues et al. 2024).

a. Refer to https://www.weforum.org/stories/2024/03/environmental-protection-cerrado-brazil-economy/.

Low water use efficiency

To assess the technical efficiency of water use at a local level, this report establishes crop-specific water use efficiency benchmarks for 35 key crops. These crops collectively accounted for more than 90 percent of global blue water consumption and virtual water trade in 2019. The benchmark values are further differentiated by production system (irrigated vs. rain fed) and climate zone (defined by the aridity index).[2] Crop-unit water footprints (m^3 per ton) were ranked within each production-climate zone pair. High-efficiency regions are identified as those where crop-unit water footprints are below the 20th percentile in their respective group; medium-efficiency regions fall between the 20th and 50th percentiles, and low-efficiency regions are those at or above the 50th percentile. Further details on the methodology can be found in technical appendix C (online).[3]

About one-third of the increase in the global water footprint has taken place in regions affected by drying.

MAP 3.1 Water footprint trends in drying regions, 2000–19

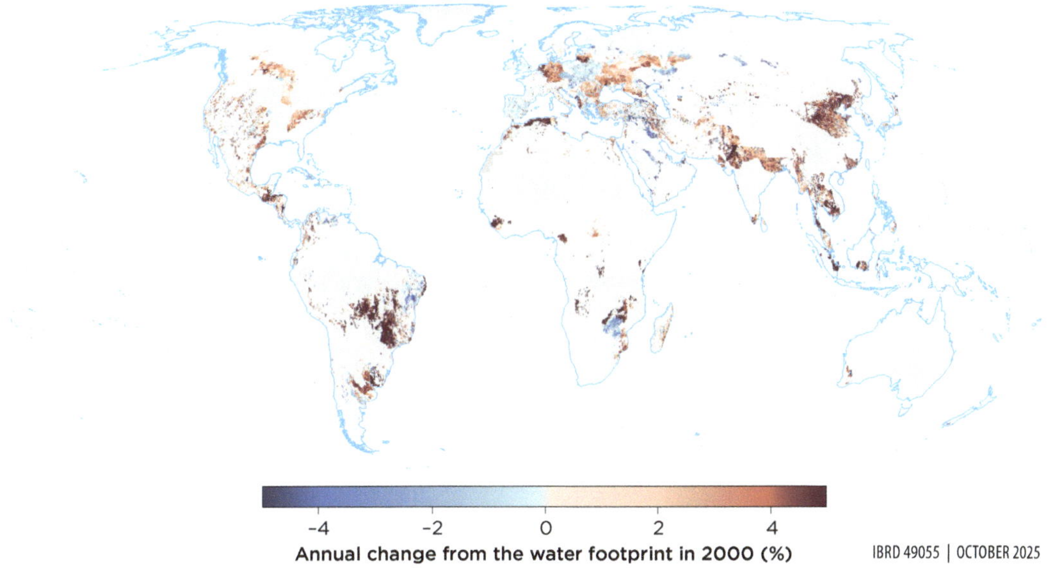

Source: World Bank.
Note: The 2000–19 trend in the water footprint is expressed as the percentage change from the local baseline (2000) footprint per year. Regions where the baseline water footprint is low (less than 10 mm) or where the terrestrial water storage trend is not declining are not included.

Map 3.2 illustrates the overlap of low crop water use efficiency and areas undergoing TWS depletion. The findings indicate that a considerable proportion of low-efficiency producers are located in drying areas. Notably, approximately one-quarter of the inefficient water footprints for rain-fed production and one-third of those for irrigated production are in regions with declining TWS. Drying hot spots with inefficient production are predominantly found in Western Asia, Eastern Europe, and North Africa, where up to 100 percent of water used with low efficiency is situated in arid areas. On a national level, the highest share of inefficient agricultural production under drying conditions is observed in Algeria, Cambodia, Mexico, Pakistan, Thailand, Tunisia, and Romania.

A large share of agriculture production in drying regions relies on inefficient water use.

MAP 3.2 Share of agricultural water use with low efficiency in drying regions

a. Rain-fed areas

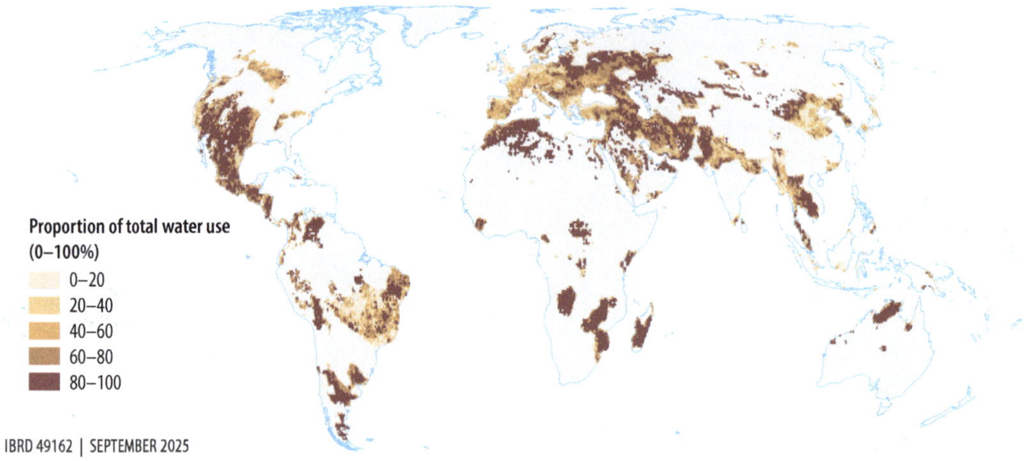

b. Irrigated areas

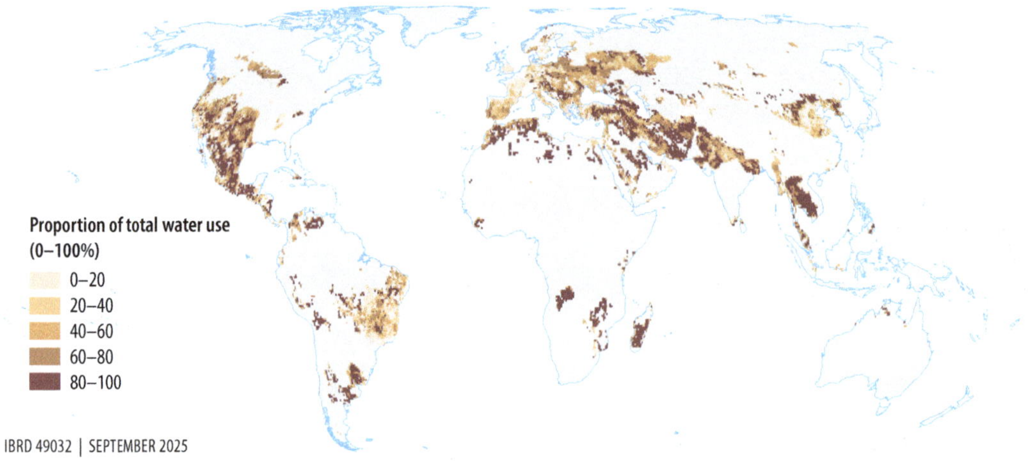

Source: World Bank.
Note: Percentages indicate the share of water use of 35 key crops cultivated with low water use efficiency in rain-fed and irrigated production systems in drying regions.

The overlap between low water use efficiency and declining TWS signals a critical vulnerability in water management. Inefficient water use may have contributed to drying in the first place, and it may continue driving depletion of already-stressed water systems as producers extract more water to compensate for inefficiencies. However, these regions are also well positioned to enhance resilience because effective solutions for improving water management can address both overuse and declining supply.

Expansion of water-intensive agriculture

Beyond water use efficiency, crop choice plays a key role in determining overall water demand. The past two decades have seen a global shift to the cultivation of more water-intensive crops. Analysis at the country level shows that 81 of the 169 countries examined have increased the water intensity of their crop production over the past two decades. Overlaying the TWS trend and the crop water intensity index further reveals that 37 countries that have shifted to more water-intensive agriculture are also facing declining freshwater reserves, including 22 countries located in arid and semi-arid regions (refer to map 3.3). Refer to the technical appendixes for details of the index decomposition analysis.

This structural shift to more water-intensive crop cultivation, coupled with inefficiency, further intensifies water demand in already water-stressed countries. Notably, more than two-thirds of the inefficient irrigation in drying areas is linked to the cultivation of water-intensive crops, including rice, wheat, cotton, maize, or sugar cane.

The choice between growing water-intensive cash crops and staple crops presents a critical trade-off for drying countries. Although cash crops can generate higher short-term profits, they often come at the cost of depleting already scarce water resources. In contrast, staple crops such as wheat, millet, and sorghum are usually less water intensive and essential for local food security but may offer lower immediate financial returns.

Although the choice between cash and staple crops depends on risk preference and local context, building long-term resilience requires channeling revenues from cash crops into skills development and human capital, especially for rural communities, to expand nonagricultural job opportunities and reduce economic reliance on an increasingly vulnerable water supply. In this way, communities can gradually shift from a water-stressed, agriculture-dependent economy to a more diversified, resilient one and be better prepared to withstand future climate and water shocks.

Many drying countries are shifting to water-intensive agriculture.

MAP 3.3 Shift to water-intensive agriculture, 2000–19

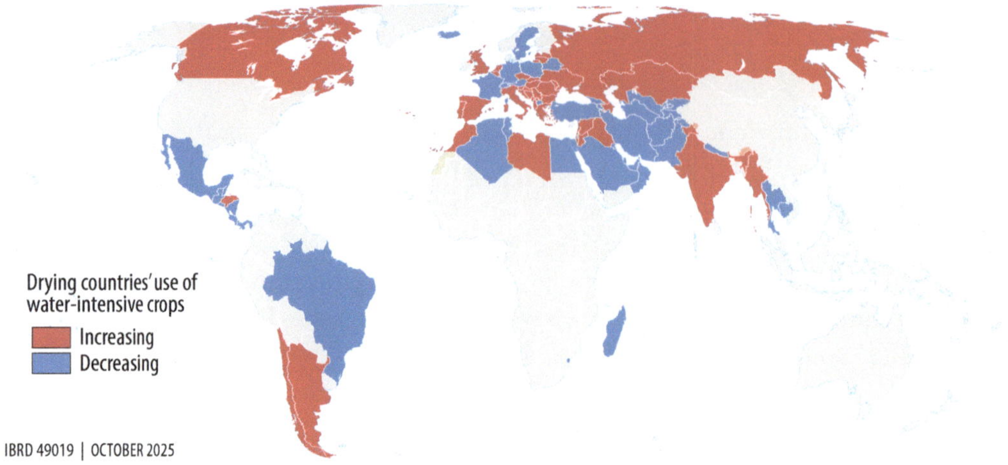

Source: World Bank.
Notes: This map denotes the trend of water intensity of crop production by country, based on an index decomposition analysis that disaggregates changes in total water intensity of crop production into changes in technical efficiency (water consumption per ton of crop) and structural change (composition of crops). Countries that grew more water-intensive crops during 2010–19 compared to 2000–10 are considered shifting to more water-intensive agriculture and are denoted in red. Blue denotes countries that are drying but shifting away from water-intensive crops based on the index decomposition analysis. Regardless of their shift to or away from water-intensive crops, countries in gray are not drying on average (even though parts of the countries can be drying). For example, the southwestern United States is a drying hot spot, but the United States on average is not drying.

Water savings potential

In response to the challenges of rising water demand and decreasing water availability, this section examines global water savings potential through two strategies: enhancing the technical efficiency of water use and improving the spatial efficiency of water allocation. The latter strategy involves reallocating the production of water-intensive crops from less efficient to more efficient areas and from regions facing water scarcity to those with abundant water supplies. This strategy can be achieved primarily through the virtual water trade and the spatial adjustment of crop production within national borders.

Improving technical efficiency

Improving crop water use efficiency has the potential to save significant amounts of water. A simulation analysis for this report indicates that the largest potential for water savings by crop can be achieved by improving the water use efficiency of maize, wheat, and rice production. The largest savings—in both absolute and relative terms—are found in South Asia (refer to map 3.4).

> **Improving crop water use efficiency has the potential to save significant amounts of water.**

MAP 3.4 National water savings potential if all inefficient producers achieved global median efficiency levels

a. Total water savings in countries with drying areas

b. Blue water savings in countries with drying areas

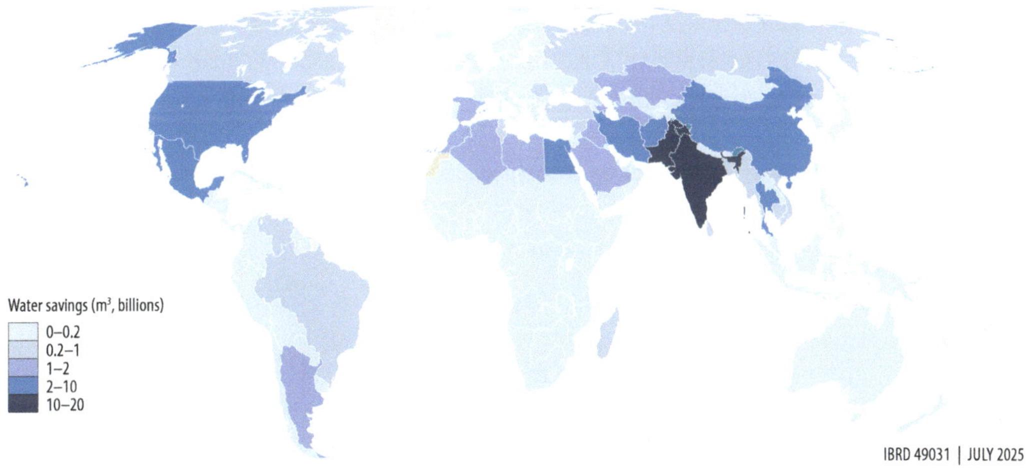

Source: World Bank.

More specifically, if all producers in drying regions currently operating at low efficiency levels—defined as below the 50th percentile of global crop water productivity—were to match the median efficiency benchmark for the 35 key crops, blue crop water consumption in drying areas would decrease by 18 percent. Further narrowing the efficiency gap by bringing crop water footprints down to the 20th percentile—reflecting the level of the most efficient production—could increase annual blue water savings in drying regions to 34 percent.

At the global level, 23 percent of the total global crop water consumption (blue and green) in the production of the 35 key crops could be saved annually if all global inefficient producers were to reach the median efficiency level in the production of the 35 key cops. Most of the savings would come from green water (942 billion m^3, representing 24 percent of the global green crop water footprint); blue water savings would account for the remainder (137 billion m^3, or 17 percent of the global blue crop water footprint, equivalent to the annual water needs of 118 million people).

If all global producers were to reach high efficiency levels, annual water savings could be increased to 36 percent of the total crop water footprint. This savings would include 1.45 trillion m^3 of green water savings (37 percent of the global green crop water footprint) and 246 billion m^3 of blue water savings (31 percent of the global blue crop water footprint).

Note that efficiency improvements alone do not guarantee real water savings. Without proper regulation, increased efficiency can sometimes lead to higher overall water consumption—a phenomenon known as Jevon's Paradox. This phenomenon occurs when efficiency gains lower the cost of water use, encouraging expanded agricultural activity or the cultivation of more water-intensive crops, ultimately intensifying water use and exacerbating scarcity (Grafton et al. 2018). To achieve sustainable water management, a balanced approach—combining technological advancements with water pricing incentives and extraction limits—is essential to ensure that improved efficiency translates into water conservation.

Improving spatial efficiency

Another strategy for reducing global water use is through optimizing virtual water trade. Although direct water trading is often difficult because of its bulkiness and challenges in transport, countries can indirectly trade water through goods that require significant water for their production. These goods—ranging from agricultural products such as grains and textiles to industrial commodities such as steel and paper—embody water, allowing water-scarce countries to import water-intensive products instead of depleting their own resources for local production. In 2019, about 25 percent of the global water footprint was traded virtually rather than consumed domestically.

Virtual water trade offers a significant opportunity to optimize global water use. By leveraging differences in water use efficiency between countries, virtual water trade can lead to global water savings. Such trade is particularly beneficial when water-scarce countries import water-intensive goods from more water-abundant countries. For instance, Jordan, a country facing significant water scarcity, saved about 7 billion m^3 of water annually through virtual water trade between 1996 and 2005 (Schyns et al. 2015).

This report estimates that crop trade contributed to global water savings of 475 billion m^3 per year between 2000 and 2019, equivalent to 9.4 percent of global water consumption of the 35 key crops. Moreover, annual water savings from virtual water trade grew by 49 percent over the past two decades, increasing from 381 billion m^3 annually between 2000 and 2009 to 569 billion m^3 between 2010 and 2019.

Whereas a few countries account for the largest share of global water savings from virtual water trade, many others achieve notable water savings relative to their consumption. From 2010 to 2019, the top 10 countries captured 49 percent of total water savings from virtual water trade, with China leading at 10 percent, followed by Indonesia at 7 percent and Japan at 6 percent. Yet, during the same period, 80 countries saved at least 10 percent of their water consumption.

An analysis of the water savings by country reveals that drying countries benefit less from the virtual water trade compared with nondrying countries. Between 2000 and 2019, drying countries saved 8.3 percent of their water consumption through virtual water trade, whereas nondrying countries saved 10.5 percent. Nondrying and nonarid countries benefited the most from virtual water trade (18.6 percent), whereas drying and nonarid countries benefited the least (4.6 percent; refer to figure 3.2).

Although virtual water trade offers the potential for water savings, it also presents challenges. Unsustainable water use can arise when exporting countries overexploit their water resources to meet the demand for virtual water export. This overexploitation can exacerbate water stress in already water-scarce regions, even contributing to water stress in water-rich countries (Rosa et al. 2019). For example, although large commercial farmers in South Africa are more efficient in crop production than those in neighboring countries, production relies on increasingly stressed water resources, which are further strained by virtual water trade (Dalin and Conway 2016).

Virtual water trade contributes to significant water savings; however, drying countries capture fewer of these benefits than nondrying countries

FIGURE 3.2 Water savings from virtual water trade, 2000–19

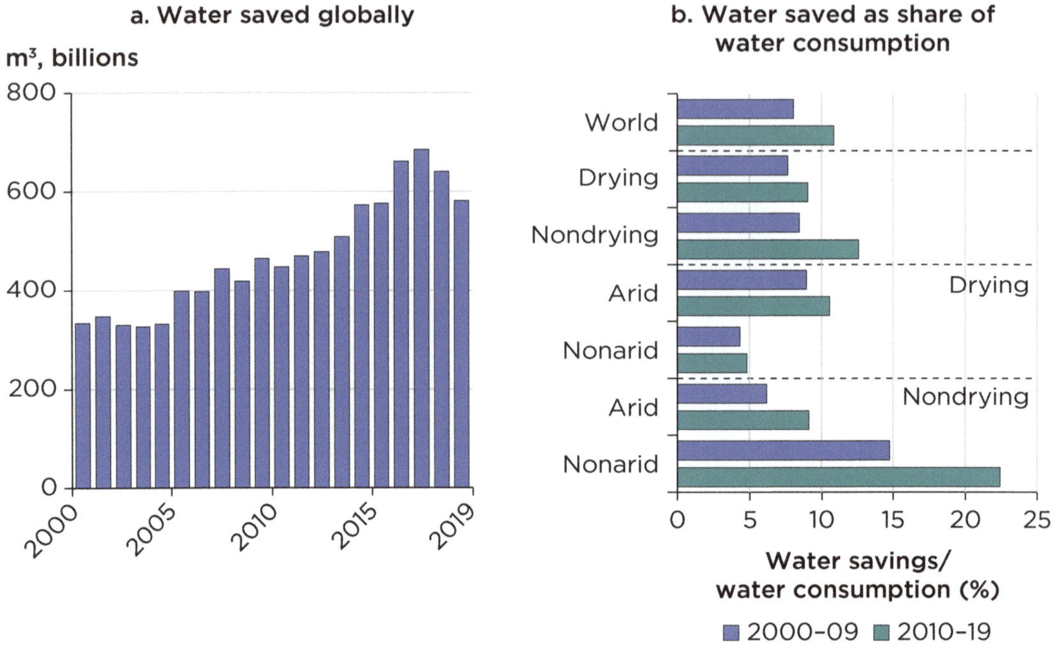

Source: World Bank.

Note: Panel a summarizes the amount of water saved when a country with a higher unit footprint for a specific crop imports products derived from this crop from a more efficient producer rather than producing the products locally. Panel b presents the decadal averages of annual water savings expressed as a share of total water consumption across country groups. *Drying* refers to countries that are drying, *arid* refers to countries that are either predominantly arid or semiarid, and *nonarid* refers to countries that are predominantly humid.

The trade of virtual water, especially blue water, has raised concerns about its sustainability. About 43 percent of global blue water trade between 1996 and 2005 was considered unsustainable, often exceeding the environmental flow requirements necessary to maintain ecosystems and biodiversity (Mekonnen and Hoekstra 2020). Similarly, even though the trade of green water is often viewed as less problematic, it can affect regions dependent on rain-fed agriculture (D'Odorico et al. 2019).

To estimate the potential to improve the efficiency and sustainability of virtual water trade, this report identifies suboptimal virtual water trade. Virtual water trade is considered suboptimal when (a) exporting countries use more water than the importer to produce the same crops, indicating no net water savings, and (b) the exporter relies more heavily on water from arid or water-stressed regions compared to the importer, which suggests the exchange may be less sustainable for the exporter. Refer to box 3.3 for details of the methodology.

BOX 3.3

Identifying inefficient and unsustainable virtual water trade

For green water, bilateral trade in a crop is considered suboptimal if (a) the exporter uses more water per ton of crop production than the importer and (b) a larger share of the exporter's crop production comes from arid areas compared with the importer. For blue water, the first condition remains the same, whereas the second is defined as a higher share of production in the exporting country coming from water-stressed areas, where blue water demand exceeds supply after accounting for environmental flow requirements.

Because of the lack of subnational-level trade data, the exact production location of traded crops within a country cannot be determined. National-level production shares from arid or water-stressed areas are therefore used to estimate the volume of suboptimal trade. Specifically, suboptimally traded water includes all inefficient and unsustainable water exported from arid (green water) or water-stressed (blue water) regions. Suboptimal virtual water trade is estimated for the 35 major crops across the 164 countries with information on both water footprint and virtual water trade.

The following example illustrates the methodology. In 2019, Mexico exported 771 million m^3 of water embedded in maize to the United States, representing 43 percent of its total maize exports. The United States required 465 m^3 of water to produce a ton of maize, whereas Mexico needed 1,021 m^3, meaning that Mexico had a lower crop water use efficiency to grow maize. Additionally, 53 percent of the water used for maize in Mexico came from arid areas, compared with 24 percent in the

(continued)

BOX 3.3
Identifying inefficient and unsustainable virtual water trade *(continued)*

United States, making the trade less sustainable for the exporter (refer to map B3.3.1). Of the total water exported, 98 percent (771 million m^3) was green water. Given that 53 percent of the production occurred in arid areas, 53 percent of the green water exports, or 409 million m^3, is regarded as suboptimally traded. This trade relationship highlights an important opportunity to improve both the efficiency and sustainability of green water use in agricultural exports.

A greater share of Mexico's maize is grown in arid areas compared to maize grown in the United States.

MAP B3.3.1 Distribution of maize crop production in the United States and Mexico

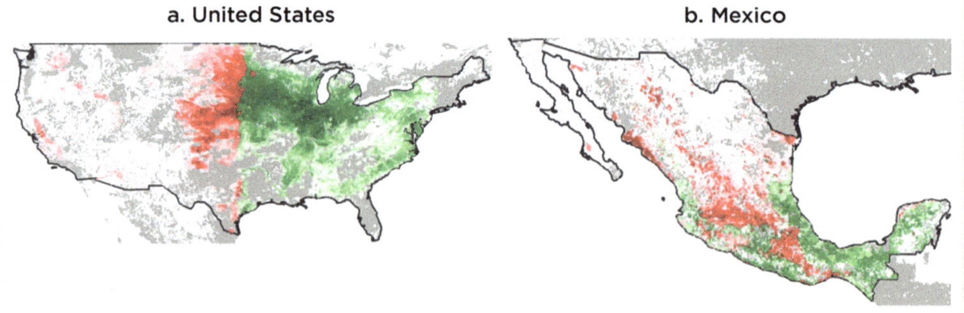

Source: World Bank.
Note: Gray areas denote cropland. Red areas denote maize production in arid areas, and green areas denote maize production in nonarid areas. The intensity of the color denotes total water use.

In 2019, about 9 percent of green water was suboptimally traded, decreasing from 14 percent in 2000. Despite the smaller overall volume of blue water virtual trade, about 21 percent of it was suboptimal. Rice and cotton were the crops with most inefficiently traded blue water (refer to figure 3.3).

The inefficiency and unsustainability of virtual water trade have been declining.

FIGURE 3.3 **Inefficient and unsustainable virtual water trade, 2000–19**

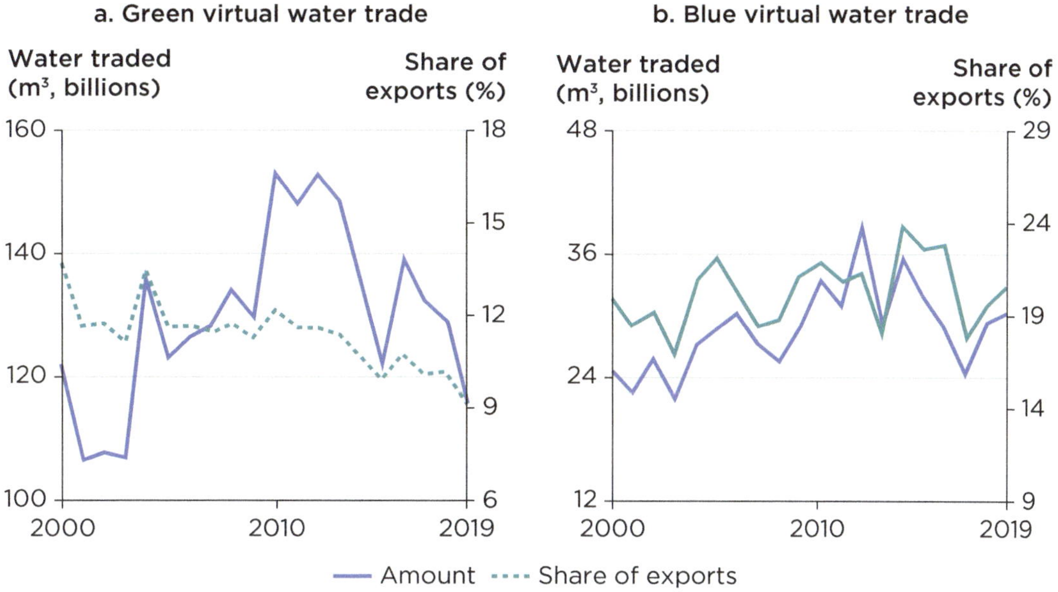

Source: World Bank.

Note: Solid lines denote the amount of water inefficiently and unsustainably traded. Dashed lines denote the amount as a proportion of the water exported.

There is considerable variation across countries in the share of suboptimal virtual water trade (refer to map 3.5). In many countries in the Middle East and North Africa, as well as in South and Central Asia—regions already facing drying conditions—inefficient and unsustainable virtual water exports account for a significant share of total virtual water exports.[4] Between 2000 and 2019, about 11 percent of virtual water export from nondrying countries was deemed suboptimal compared with 17 percent of such exports from drying and arid countries. The share of suboptimal water trade embedded in irrigated crops from drying and arid countries was even higher, reaching 27 percent over the same period.

Many water-scarce countries engage in suboptimal virtual water export.

MAP 3.5 Share of suboptimal virtual water trade, by country of origin, 2000–19

a. Green water

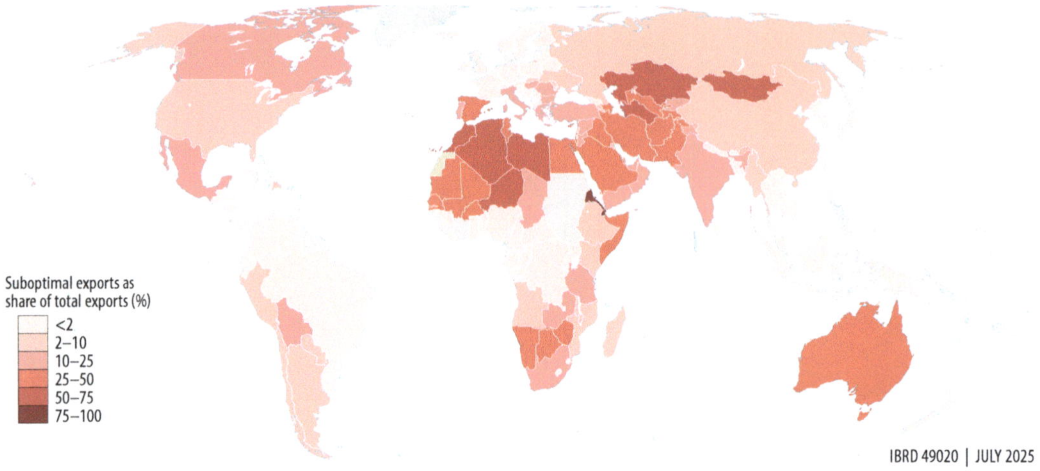

b. Blue water

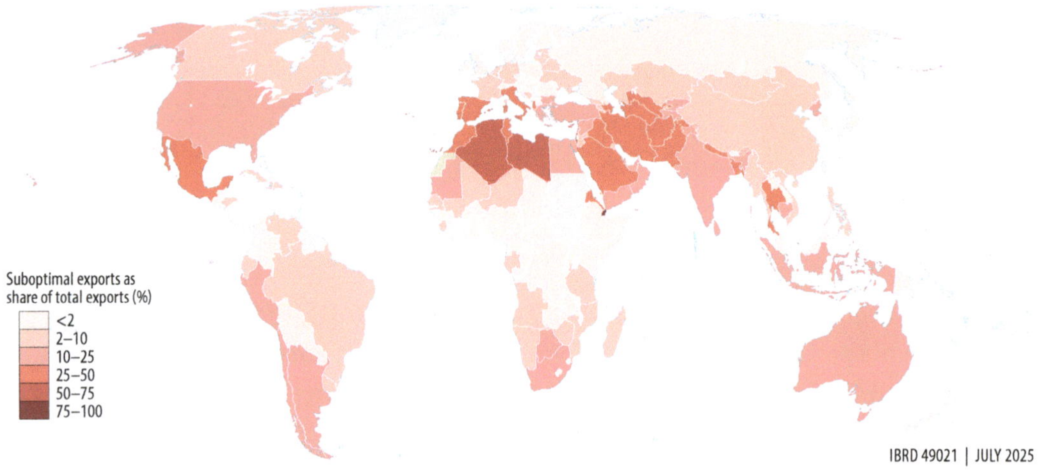

Source: World Bank.
Note: Countries shaded in gray export little to no water. Increasing color intensity indicates increasing share of suboptimal virtual water export as a share of total exports.

This analysis suggests that, although virtual water trade has overall contributed to global water savings, improving efficiency and sustainability in the current system could further enhance the gain. Although trade outcomes are influenced by various factors of comparative advantage, such as capital and labor (Bombardini,

Gallipoli, and Pupato 2012; Romalis 2004), institutions (Levchenko 2007; Nunn 2007), and demographic composition (Cai and Stoyanov 2016), studies have shown that water also plays a crucial role, especially in water-intensive sectors (Carleton, Crews, and Nath 2025; Debaere 2014; Liu et al. 2024). However, distortions in the agricultural input and output market—such as energy subsidies for groundwater pumping and price guarantees for certain crops—can reinforce comparative disadvantage. These policies may hinder structural transformation and lead water-scarce countries to continue producing and exporting water-intensive products, further straining their already limited resources (Damania et al. 2023). Agricultural production must balance multiple objectives, such as food security, economic growth, rural development, and jobs creation; however, prioritizing efficiency and savings by removing pricing distortions can lead to more sustainable outcomes and maximize the benefits from global virtual water trade.

Another key strategy to improve the efficiency of the virtual water trade is to enhance the efficiency of water use among the least efficient producers and to adjust cropland distribution to better align with water availability. Cropland redistribution, in particular, can be very promising in terms of reducing water footprints. A study finds that optimizing global cropland distribution could remove the need for irrigation altogether, allowing rain-fed agriculture to sustain current yields (Beyer et al. 2022). Nonetheless, given the difficulty of spatial relocation of croplands, countries can also opt for better crop redistribution within existing croplands. Such a redistribution is estimated to generate water savings of at least 20 percent for many water-stressed countries (Davis et al. 2017).

Priority regions

Countries face wide variation in the severity and nature of their water challenges: some are driven by supply-side constraints, others by rising demand, and many by a combination of both. To help identify priority regions, two composite indexes are developed to reflect supply and demand stressors at the country level. The supply-side index measures the rate of water depletion and baseline water endowment (aridity). The demand-side index measures crop water use efficiency, water intensity of crops, and share of suboptimal virtual water export. The placement of each region corresponds to the population-weighted mean of the index across all countries within the region (refer to figure 3.4).

Priority regions can be identified based on supply- and demand-side stressors.

FIGURE 3.4 **Composite indexes measuring supply- and demand-side stressors**

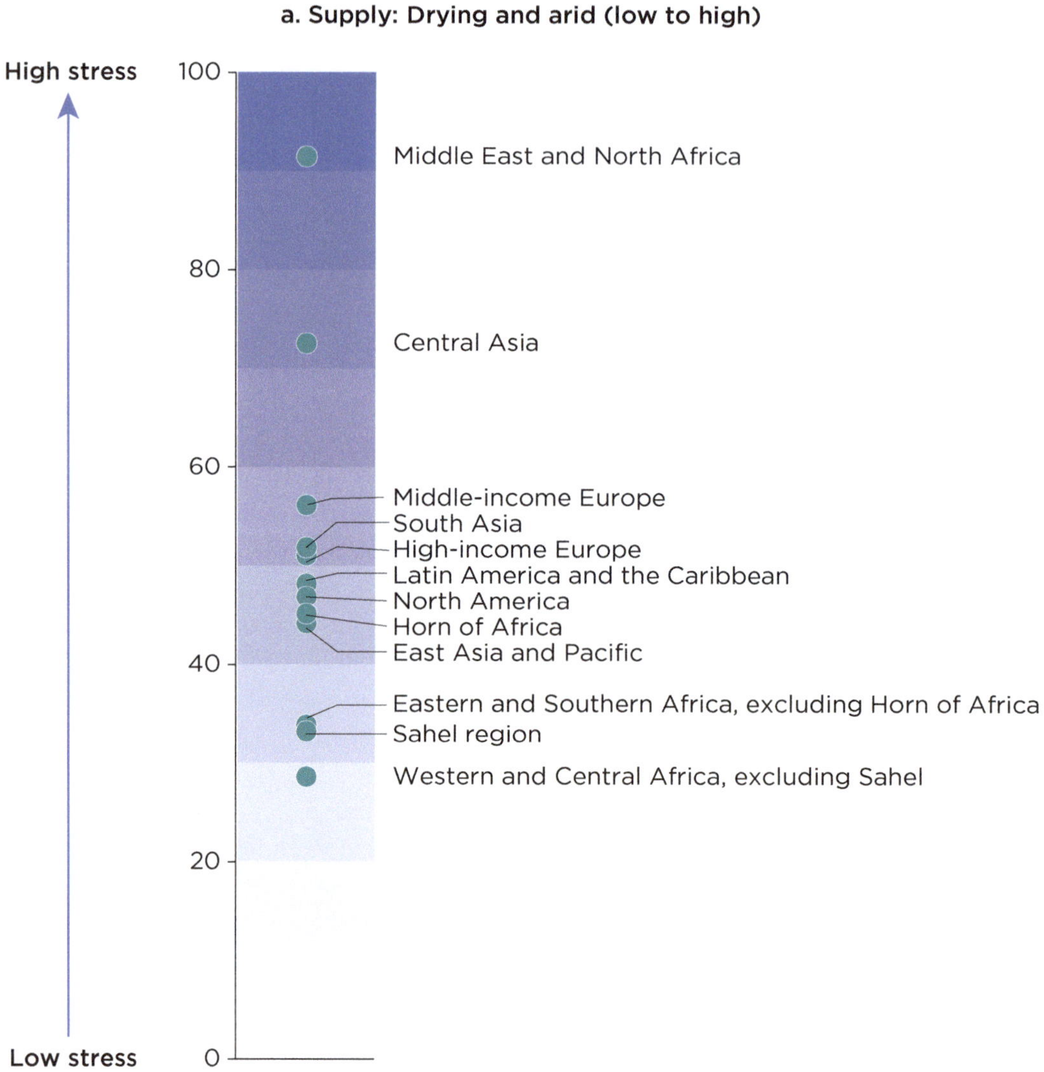

(continued)

FIGURE 3.4 Composite indexes measuring supply- and demand-side stressors *(continued)*

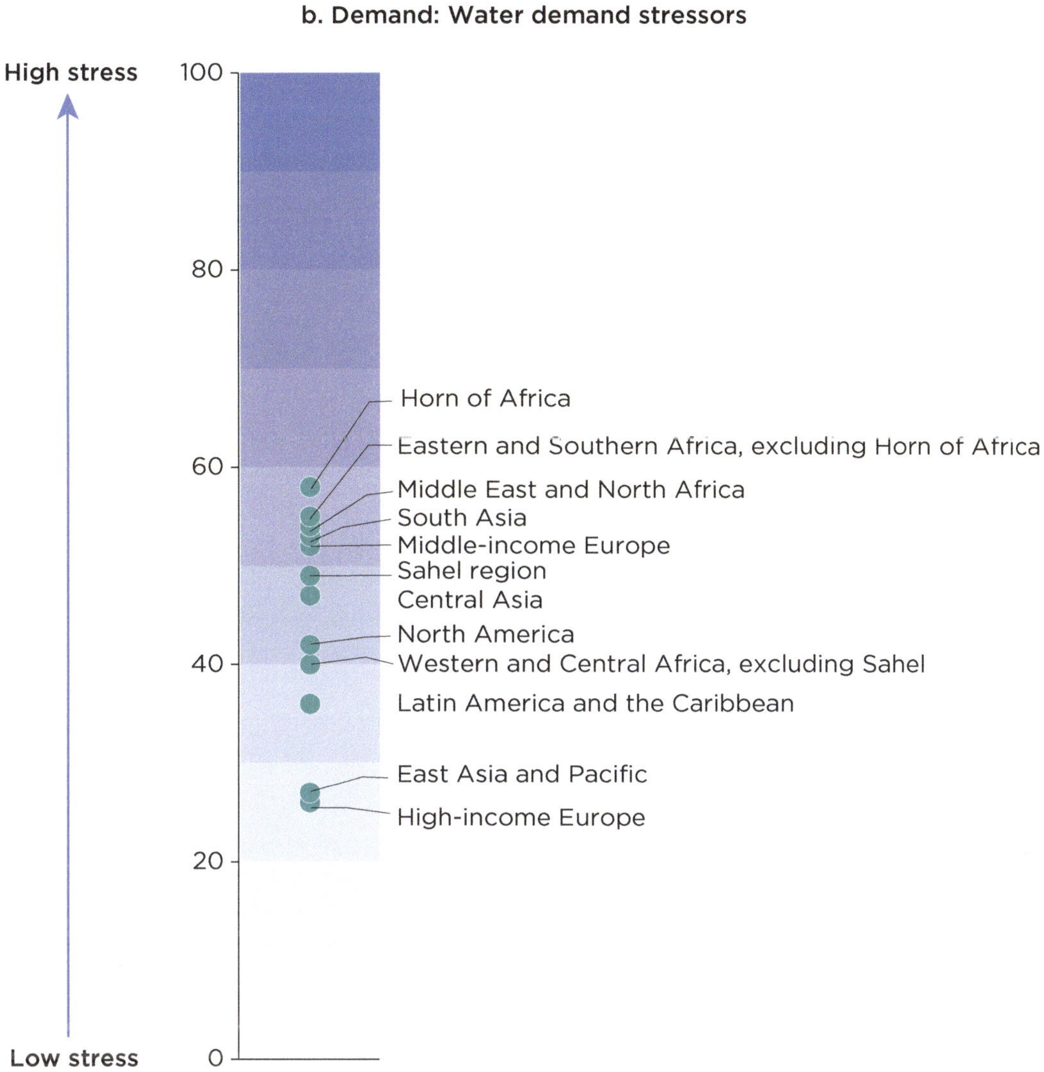

Source: World Bank
Notes: The supply index considers how fast countries are drying as well as their aridity. The demand index focuses on crops and considers efficiency and water intensity in production as well as efficiency in virtual water trade. The placement of each region corresponds to the population-weighted mean of the index across all countries within the region.

Note that regional averages can mask significant variations at the country level. For example, in Sub-Saharan Africa, countries such as the Democratic Republic of Congo do not face physical water supply constraints, whereas countries such as Chad, Mauritania, Niger, and the Federal Republic of Somalia do. Moreover, in large countries such as China, Mexico, and the United States, national averages may overlook important subnational differences. For example, northern China and the southwestern United States are already facing severe water stress and are part of the mega-drying regions identified in map 1.1.

Conclusion

This chapter estimates water consumption patterns, focusing on the agricultural sector, to identify hot spot regions most vulnerable to continental drying. These regions are characterized by inefficient crop water use, expansion of water-intensive crops, and a high share of inefficient and unsustainable virtual water export. However, there are significant opportunities to reduce these inefficiencies, particularly by enhancing technical and spatial water use efficiency—key steps toward improving resilience.

Notes

1. The 23 percent share of livestock in the total water footprint does not consider contribution from crops grown for multiple purposes, including livestock feed (e.g., maize and soybean). Hence, the actual share of livestock in the total water footprint is likely underestimated.
2. Crop water use efficiency is influenced not only by agricultural practices and technologies but also by environmental factors, particularly climate (Zhuo, Mekonnen, and Hoekstra 2016). Typically, arid regions require more water per ton of produce than humid regions.
3. Technical appendixes A through E are available online at https://hdl.handle.net/10986/43683.
4. One caveat of this analysis is that some water-scarce countries use desalinated water to produce and export agricultural goods, meaning the water is not drawn from freshwater sources. However, current data do not allow for estimation of the share of desalinated versus freshwater used in these virtual water exports or for assessment of the opportunity cost, especially because desalinated water could serve other sectors. In countries in which a significant share of agricultural water comes from economically viable desalination, this chapter's estimates of unsustainable trade may be overstated.

References

Beyer, R. M., F. Hua, P. A. Martin, A. Manica, and T. Rademacher. 2022. "Relocating Croplands Could Drastically Reduce the Environmental Impacts of Global Food Production." *Communications Earth & Environment* 3 (1): 1–11. https://doi.org/10.1038/s43247-022-00360-6.

Bombardini, M., G. Gallipoli, and C. Pupato. 2012. "Skill Dispersion and Trade Flows." *American Economic Review* 102 (5): 2327–48. https://doi.org/10.1257/aer.102.5.2327.

Cai, J., and A. Stoyanov. 2016. "Population Aging and Comparative Advantage." *Journal of International Economics* 102: 1–21. https://doi.org/10.1016/j.jinteco.2016.04.006.

Carleton, Tamma, Levi Crews, and Ishan Nath. 2025. "Agriculture, Trade, and the Spatial Efficiency of Global Water Use." Unpublished manuscript, last modified April 11, 2025. https://www.levicrews.com/files/p-wateruse_paper.pdf.

Dalin, C., and D. Conway. 2016. "Water Resources Transfers through Southern African Food Trade: Water Efficiency and Climate Signals." *Environmental Research Letters* 11 (1): 015005. https://doi.org/10.1088/1748-9326/11/1/015005.

Damania, Richard, Esteban Balseca, Charlotte de Fontaubert, Joshua Gill, Kichan Kim, Jun Rentschler, Jason Russ, and Esha Zaveri. 2023. *Detox Development: Repurposing Environmentally Harmful Subsidies*. Washington, DC: World Bank.

Davis, K. F., M. C. Rulli, A. Seveso, and P. D'Odorico. 2017. "Increased Food Production and Reduced Water Use through Optimized Crop Distribution." *Nature Geoscience* 10 (12): 919–24. https://doi.org/10.1038/s41561-017-0004-5.

Debaere, P. 2014. "The Global Economics of Water: Is Water a Source of Comparative Advantage?" *American Economic Journal: Applied Economics* 6 (2): 32–48. https://doi.org/10.1257/app.6.2.32.

D'Odorico, P., J. Carr, C. Dalin, J. Dell'Angelo, M. Konar, F. Laio, L. Ridolfi, et al. 2019. "Global Virtual Water Trade and the Hydrological Cycle: Patterns, Drivers, and Socio-Environmental Impacts." *Environmental Research Letters* 14 (5): 053001. https://doi.org/10.1088/1748-9326/ab05f4.

Grafton, R. Q., J. Williams, C. J. Perry, F. Molle, C. Ringler, P. Steduto, B. Udall, et al. 2018. "The Paradox of Irrigation Efficiency." *Nature Sustainability* 1 (9): 466–72. https://doi.org/10.1038/s41893-018-0138-5.

Hoekstra, A. Y., and M. M. Mekonnen. 2012. "The Water Footprint of Humanity." *Proceedings of the National Academy of Sciences* 109 (9): 3232–7. https://doi.org/10.1073/pnas.1109936109.

Latrubesse, E. M., E. Arima, M. E. Ferreira, S. H. Nogueira, F. Wittmann, M. S. Dias, F. C. P. Dagosta, and M. Beyer. 2019. "Fostering Water Resource Governance and Conservation in the Brazilian Cerrado Biome." *Conservation Science and Practice*, 1: e77. https://doi.org/10.1111/csp2.77.

Levchenko, A. A. 2007. "Institutional Quality and International Trade." *Review of Economic Studies* 74 (3): 791–819. https://doi.org/10.1111/j.1467-937X.2007.00435.x.

Liu, X., Y. Xu, S. Sun, P. Wu, and Y. Wang. 2024. "Water Resource Efficiency Evaluation of Crop Production in Arid and Semi-Arid Regions Based on Water Footprint and Comparative Advantage." *European Journal of Agronomy 160*: 127310. https://doi.org/10.1016/j.eja.2024.127310.

Mekonnen, M. M., and A. Y. Hoekstra. 2020. "Sustainability of the Blue Water Footprint of Crops." *Advances in Water Resources* 143: 103679. https://doi.org/10.1016/j.advwatres.2020.103679.

Mialyk, O., J. F. Schyns, M. J. Booij, H. Su, R. J. Hogeboom, and M. Berger. 2024. "Water Footprints and Crop Water Use of 175 Individual Crops for 1990–2019 Simulated with a Global Crop Model." *Scientific Data* 11 (1): 206. https://doi.org/10.1038/s41597-024-03051-3. Background paper for this report.

Nunn, N. 2007. "Relationship-Specificity, Incomplete Contracts, and the Pattern of Trade." *Quarterly Journal of Economics* 122 (2): 569–600. https://doi.org/10.1162/qjec.122.2.569.

Rodrigues, A. A., M. N. Macedo, D. V. Silvério, L. Maracahipes, M. T. Coe, P. M. Brando, J. Z. Shimbo, R. Rajão, B. Soares-Filho, and M. M. C. Bustamante. 2022. "Cerrado Deforestation Threatens Regional Climate and Water Availability for Agriculture and Ecosystems." *Global Change Biology*, 28: 6807–22. https://doi.org/10.1111/gcb.16386.

Romalis, J. 2004. "Factor Proportions and the Structure of Commodity Trade." *American Economic Review* 94 (1): 67–97. https://doi.org/10.1257/000282804322970715.

Rosa, L., D. D. Chiarelli, C. Tu, M. C. Rulli, and P. D'Odorico. 2019. "Global Unsustainable Virtual Water Flows in Agricultural Trade." *Environmental Research Letters* 14 (11): 114001. https://doi.org/10.1088/1748-9326/ab4bfc.

Schyns, J. F., A. Hamaideh, A. Y. Hoekstra, M. M. Mekonnen, and M. Schyns. 2015. "Mitigating the Risk of Extreme Water Scarcity and Dependency: The Case of Jordan." *Water* 7 (10): 5705–30.

Spera, S. A., G. L. Galford, M. T. Coe, M. N. Macedo, and J. F. Mustard. 2016. "Land-Use Change Affects Water Recycling in Brazil's Last Agricultural Frontier." *Global Change Biology*, 22: 3405–13. https://onlinelibrary.wiley.com/doi/10.1111/gcb.13298.

Zhuo, L., M. M. Mekonnen, and A. Y. Hoekstra. 2016. "Benchmark Levels for the Consumptive Water Footprint of Crop Production for Different Environmental Conditions: A Case Study for Winter Wheat in China." *Hydrology and Earth System Sciences* 20 (11): 4547–59. https://doi.org/10.5194/hess-20-4547-2016.

CHAPTER 4

Policy Recommendations: Solutions to the Continental Drying Crisis

Key messages

- This chapter outlines a three-pronged approach to address and adapt to the challenges posed by continental drying. First, prioritize effective water demand-side management by adopting water-efficient technologies, establishing water abstraction limits, and implementing public awareness campaigns to encourage behavior change. Second, augment water supply through measures such as water recycling and reuse, desalination, and water storage. Third, improve water allocation to ensure that scarce water resources are distributed fairly and efficiently in an increasingly drying world.
- Five cross-cutting levers are needed to create an enabling environment for the effective implementation of this approach. These levers are strengthening institutions, reforming tariffs and repurposing subsidies, adopting water accounting, leveraging data and technological innovations, and valuing water in trade. These levers are also critical for mobilizing private finances and expertise for sustainable water management.
- The framework presented in this chapter outlines actions in the water sector, with a particular emphasis on the agriculture sector, the largest water-consuming sector globally. However, as highlighted throughout this report, addressing trade barriers, investing in education and skills development, and improving access to markets and financial services are equally critical for strengthening job and livelihood resilience amid a continental drying crisis.

A reproducibility package is available for this book in the Reproducible Research Repository at https://reproducibility.worldbank.org/index.php/catalog/386.

Introduction

This chapter lays out a policy framework to address the escalating water crisis marked by continental drying, rising water demand, and inadequate water management, with a particular focus on the agricultural sector, the largest water user globally. Addressing the growing challenge of continental drying requires a deeply integrated and cooperative approach to water management. As illustrated in figure 4.1, at the core of the framework lie three policy pillars.

The first pillar focuses on managing water demand. As highlighted in chapter 3, there is widespread inefficiency in water consumption and great potential for water savings by eliminating that inefficiency. Managing water demand requires the adoption of water-efficient technologies and water extraction limits to curb overuse, as well as public awareness campaigns to promote behavioral change.

The second pillar focuses on increasing water supply through strategies such as water recycling, desalination, and restoring and maintaining natural and built water storage, including measures such as rainwater harvesting and groundwater recharge.

The third pillar involves improving water allocation to ensure scarce water resources are distributed fairly and efficiently. Effective water allocation—designed to determine who gets how much water, when, and for what purpose—is the backbone of sustainable water management. Water allocation becomes more critical than ever in the face of growing competition among agriculture, industry, domestic needs, and ecosystems in a drying world. Effective allocation also depends on the ability to reallocate water as supply and demand fluctuate, relying on a mixture of administrative procedures, judicial rulings, water user associations, and market-based mechanisms (Marston and Cai 2016; Meinzen-Dick and Ringler 2008; O'Donnell and Garrick 2019).

The three pillars of action—demand management, supply augmentation, and water allocation—are mutually reinforcing and interdependent. Demand management is often a prerequisite for successful supply-side interventions and water allocation reforms. Conversely, improving supply without corresponding improvements in demand management is ultimately unsustainable and risks exacerbating current water stress. Water allocation, as a core component of demand management, must be pursued alongside efforts to effectively manage both supply and demand. Much like a three-legged stool, the system cannot remain stable if any one element is absent.

A proposed policy framework for tackling continental drying includes three pillars and five policy levers.

FIGURE 4.1 Policy framework to address the continental drying crisis

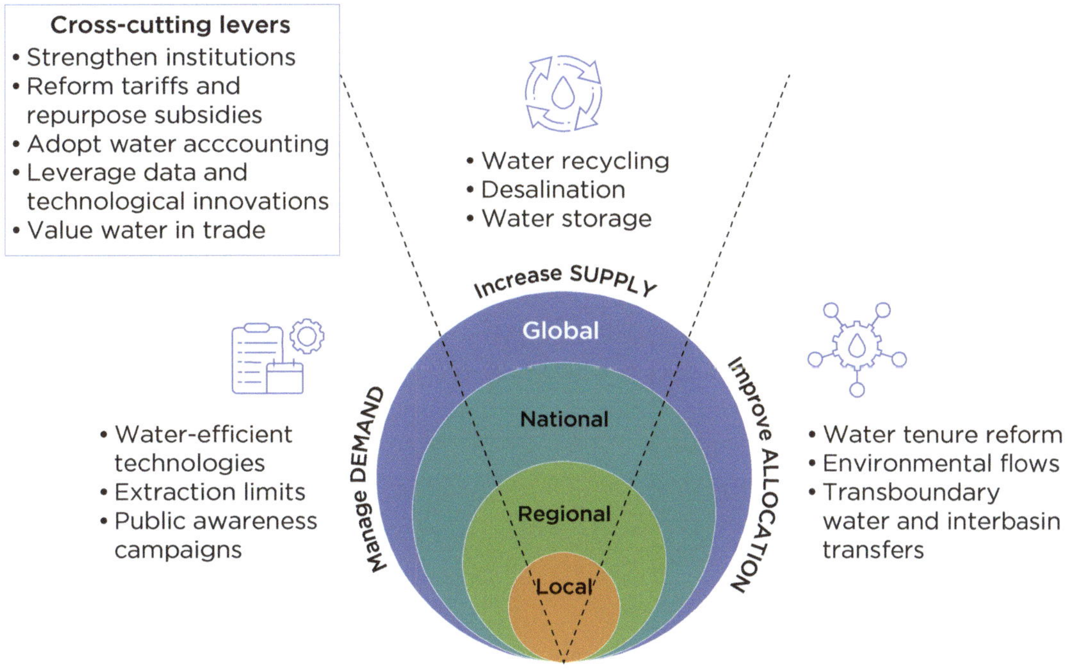

Source: World Bank.
Note: Figure schematic adapted from Molle (2003).

Five cross-cutting levers enhance the effectiveness of the three pillars: strengthening institutions, reforming water tariffs and repurposing subsidies, adopting water accounting, leveraging data and technological innovations, and valuing water in trade.

As discussed in chapter 1, effective water resource management requires appropriate land stewardship, involving cross-sectoral approaches with agriculture, forestry, and urban planning to reshape land management practices affecting water availability. The water sector can influence broader drivers through improved green water management practices that connect soil moisture, vegetation, and land cover to hydrological outcomes.

Bridging the global water investment gap involves unlocking private capital through institutional reforms, legal frameworks, and innovative financing tools, such as blended finance and green bonds.

Finally, as highlighted throughout this report, addressing trade barriers, investing in education, and improving market access are crucial for building adaptive and resilient economies that can withstand water scarcity impacts. Strategic trade relationships with less climate-vulnerable and more water-abundant countries can improve resilience, and investing in human capital and market connectivity supports rural nonfarm economic growth and adaptation to changing labor market demands.

Managing demand

This section outlines three strategies for managing demand: promoting the adoption of water-efficient technologies, setting water extraction limits, and raising public awareness through education campaigns to encourage water conservation. Water pricing creates incentives for the adoption of water-efficient technologies and is discussed as a cross-cutting policy lever.

Water-efficient technologies

In agriculture, advanced technologies, such as big data analytics, smart irrigation, and precision sprinklers, can greatly reduce water waste while keeping crops healthy. Effective use of these technologies requires careful regulation and monitoring. Tools such as remote sensing can facilitate water accounting and realize actual water savings rather than just reducing water withdrawals (refer to box 4.1).

Although this section focuses on improving agricultural water productivity, significant inefficiencies exist in water supply and distribution. Developing countries lose a large volume of water as a result of nonrevenue water (NRW). NRW reduction programs by water utilities address physical losses such as pipe leaks and commercial losses from metering errors or unauthorized use. Although some losses return to the system, reducing NRW is crucial for resilience in drought-prone regions and areas with declining water availability. For instance, investments in NRW reduction in Faisalabad, Pakistan, improved service delivery, increased utility revenues, and promoted efficient water use across the urban network (Ogata et al. 2021).

Drip, center-pivot, and sprinkler irrigation

The global adoption of drip, center-pivot and sprinkler irrigation is increasing, although not as rapidly as desired. In Saudi Arabia, two-thirds of irrigated areas use center-pivot systems, turning arid lands into productive zones for high-value crops. In the United States, Nebraska has shifted from

BOX 4.1
Real versus apparent irrigation water savings

Improving irrigation efficiency goes beyond field- and scheme-level upgrades. Classical efficiency metrics often ignore downstream water reuse, groundwater recharge, and farmers' responses to production constraints. Overirrigation can result from unreliable water supplies, weed control, or aquifer recharge needs. Advanced technologies and new efficiency metrics are vital for addressing these issues.

Water accounting tools are crucial for tracking local water supply and demand, especially where basin-scale data are lacking. These tools differentiate between real water savings, which reduce consumption and nonrecoverable flows, and apparent savings, which lower withdrawals without changing consumption. Traditional irrigation efficiency metrics often promote apparent water savings by reducing withdrawals but do not address actual consumption. Real water savings focus on evapotranspiration and nonrecoverable outflows, such as water lost to saline aquifers or oceans. These real savings free up water for other uses and are considered effective for sustainable water management.

dryland farming to irrigated corn production using these systems (Mitchell et al. 2016). China's North Plains have seen improvements in water management with center-pivot irrigation in dry, hot wind conditions (Cai et al. 2022). In Ghana, solar-powered center-pivot systems have increased vegetable yields and profits by 13 percent and 8 percent, respectively, where utilized (Baidoo et al. 2024). This technology has boosted crop yields and supported sustainable water use, reducing agricultural water withdrawals by nearly 7 billion cubic meters annually over the past decade (Al-Ghobari and Dewidar 2021; Pfeiffer and Lin 2014).

Drip irrigation has helped small farmers adapt to climate change and scarce rainfall. In Afghanistan, it supports chili pepper cultivation in arid regions with inconsistent water supplies (Hussainzada and Lee 2022). Drip irrigation from the Balkhab River allows extra downstream water usage without upstream effects, avoiding the need for a dam. In Pakistan, inexpensive drip systems have boosted yields by 20 percent compared with traditional methods (Aziz et al. 2021). Drip irrigation has also increased agricultural productivity and reduced poverty among small farmers in Ghana and Rwanda (Gaspard et al. 2023). In regions such as Minas Gerais and northeast Brazil, drip systems reduce water loss from evaporation. These systems support federal policies for water conservation and higher crop yields (Alcoforado de Moraes et al. 2021; Almeida et al. 2017).

The implementation of these irrigation technologies necessitates substantial initial investment for installation, along with continuous maintenance and energy consumption. Although government subsidies, microloans, and international funding are often essential, they remain inadequate for widespread adoption in low-income countries. Therefore, it is crucial to mobilize both private and public capital for the adoption of this approach.

Alternate wetting and drying, and system of rice intensification

Water-saving techniques such as alternate wetting and drying (AWD) and the system of rice intensification (SRI) are increasingly used in water-stressed countries to improve water use efficiency, reduce methane emissions, increase productivity, and lower arsenic levels in rice grains. Promoted by the International Rice Research Institute and local programs, AWD can save up to 38 percent of water while maintaining rice yields similar to traditional methods (Alauddin et al. 2020; Neogi et al. 2018). When the irrigation system is able to control water delivery, these techniques can reduce the need for continuous flooding in rice production. Moreover, these methods can help meet international trade standards for crops such as rice by improving irrigation efficiency. In Bangladesh, AWD has bolstered food exports; in the Philippines, AWD is part of rice production and export strategies. Studies show that AWD can cut water use by 30 percent without affecting yields, addressing regional water scarcity (Dahlgreen and Parr 2024; Nelson et al. 2015).[1]

SRI, developed in Madagascar in the 1980s, has been successful in increasing rice yields and reducing water use. In Madagascar, SRI has led to yield increases of 50 percent with less water usage (Barrett et al. 2021). In Indonesia, SRI has resulted in better water management and yield improvements, cutting water use by 50 percent (Arsil et al. 2022; Dwipa et al. 2020). The Indonesian government supports SRI through training programs and demonstration plots to promote sustainable rice production.

Agronomic improvements

Advanced agricultural technologies and agronomic innovations can reduce the demand for irrigation water and improve water productivity. These innovations include selecting drought-resistant crops and varieties, applying deficit irrigation techniques, leveraging genetic improvements in crop drought tolerance, and adjusting the cropping calendar.

Choosing water-efficient or drought-tolerant crops can significantly decrease irrigation requirements while maintaining farm profitability. For instance, shifting to less water-intensive crop mixes has notable water-savings potential. A recent multibasin study in the western United States found that changing crop patterns, such as reducing the cultivation of water-intensive alfalfa or hay in favor of alternative crops combined with some land fallowing, could lower agricultural water use by 28–57 percent across six water-scarce river basins (Richter et al. 2023). Similarly, in heavily irrigated rice-growing areas of northern India, replacing approximately one-third of rice acreage with other cereals and pulses is projected to save between 61 and 108 km^3 of groundwater under future climate scenarios, compared with continued rice cultivation (Dangar and Mishra 2024).

Certain crops naturally require less water. In Texas, grain sorghum uses about 195 mm less water per season than maize. Sorghum and millet are more drought-resistant than maize or rice and yield well in dry conditions. Farmers can reduce irrigation by choosing crops suitable for arid climates or drought-tolerant varieties of the same crop. Improving crop genetics through conventional breeding, marker-assisted selection, and biotechnology is vital in different climates. Agricultural policies support developing drought-resistant, water-efficient cultivars to enhance yield per unit of water or maintain yields under water-limited conditions.

Agronomic improvements must align with the cropping calendar to match periods of higher rainfall, reducing the need for supplemental irrigation. This matching means planting crops when they benefit most from precipitation and scheduling to avoid peak evapotranspiration. Farmers can use deficit irrigation, supplying less water than needed, to conserve water with minimal yield impact. Knowledgeable application of this technique significantly boosts water use efficiency. Properly managed deficit irrigation can save substantial water without reducing yields (FAO 2000). Countries should include these strategies in agricultural plans to optimize water use during scarcity.

Extraction limits

Improving water use efficiency alone does not ensure water savings. Efficiency gains can lower costs and lead to a rebound effect: increasing the value of each drop applied to crops can drive expansion of croplands or intensification of agricultural activity, possibly worsening water scarcity (Grafton et al. 2018). To conserve water, many countries have set extraction limits tailored to risks and capacity, ranging from moratoria and quotas to volumetric restrictions (Wight et al. 2025).

Countries regulate surface water withdrawals from rivers, lakes, and reservoirs to manage usage. Growing awareness of aquifer depletion has led some governments to limit groundwater extraction.

Basin-level planning is essential for setting water extraction limits. Countries may benefit from adopting a holistic, integrated approach to water management. This approach involves combining legal and regulatory frameworks for surface and groundwater, rather than managing them separately, and progressively addressing the full complexity of the water cycle.

Strong water institutions are crucial for ensuring that water users comply with water limit regulations (refer to the "Strengthen institutions" section later in this chapter). Recent regulatory reforms in South Africa have focused on protecting groundwater resources during fracking operations related to unconventional oil and gas extraction, which require substantial water volumes. These regulations include well spacing, water quality discharge standards, and limits on the volume of water that can be extracted. They aim to minimize impacts on groundwater resources (Esterhuyse, Vermeulen, and Glazewski 2019; Worthmann and Esterhuyse 2022). Enforcing these regulations is critical because South Africa's water resources are heavily allocated, making effective management essential to prevent overextraction and contamination. Some urban centers have benefited from increased access to water because of these regulations.

Chinese water management regulations focus on climate change adaptation and reducing water demand. The framework includes strict water extraction limits, efficiency measures, and savings options, especially in agriculture (Ashofteh et al. 2019). In regions facing water scarcity, local governments have implemented practices such as rainwater harvesting and wastewater reclamation to adhere to these limits. These regulations also aim to recharge groundwater and promote efficient irrigation, which helps to reduce nitrate pollution (Du et al. 2024). California mandates water restrictions during droughts and limits groundwater extraction to reduce consumption (Green et al. 2017).

Ethiopia's government introduced temporary policies to manage residential water demand, including extraction limits for urban water systems and conservation campaigns. These strategies have led to as much as a 15 percent reduction in water usage (Timotewos and Barjenbruch 2024; Timotewos, Barjenbruch, and Behailu 2023). The Jucar River Basin in Spain illustrates the enforcement of extraction limits to balance water supply and demand. Stakeholders in the basin have created adaptation strategies and innovations, such as modifying irrigation practices and using nonconventional water resources, to adhere to established extraction limits. This approach has led to better water management and groundwater conservation across the basin (Marcos-Garcia et al. 2023).

Water extraction limits are often applied during periods of temporary water scarcity (Li et al. 2021). If these measures are not accompanied by changes in behavior toward more conscious water use, water demand may increase and return to previous levels once restrictions are lifted. The next section explains how public awareness campaigns contribute to promoting long-term behavioral change.

Public awareness campaigns and education

Demand-side water management uses awareness campaigns to change consumption behavior and encourage water stewardship. Countries such as Jordan and South Africa have implemented such programs with significant success. Jordan's campaign aligns public concerns with government strategies, promoting sustainable water use (Benedict and Hussein 2019). Conservation campaigns can reduce water usage by up to 15 percent (Timotewos and Barjenbruch 2024; Timotewos, Barjenbruch, and Behailu 2023). During Cape Town's "Day Zero" crisis, smart meter feedback, school competitions, and strong communications campaigns reduced water use by 15–26 percent (Brühl and Visser 2021). These examples highlight the effectiveness of well-designed informational strategies in demand-side management.

Another significant attribute of behavioral change programs is their nonprice methodology, which serves as a viable alternative to increasing water rates—a politically sensitive action. In the United States, numerous water utilities have established programs aimed at reducing water consumption, with a focus on nonessential activities such as ornamental landscaping. These initiatives typically involve home water reports that compare household usage with that of neighboring households, provide technical guidance (such as methods for detecting leaks), and emphasize prosocial values such as environmental stewardship (Ferraro and Price 2013; Stitzel and Rogers 2023). The integration of technical advice with prosocial messaging has proven effective in substantially lowering water usage and fostering enduring behavioral modifications (Brent et al. 2020; Ferraro and Price 2013).

Providing information is crucial for farmers and rural communities facing climate-related water risks. Like residential users, farmers respond to information and social cues (Peth et al. 2018). Many farmers are unaware of the benefits of water-efficient innovations (Levidow et al. 2014). High costs, limited knowledge, and lack of data are major barriers (Jararweh et al. 2023; Mi et al. 2021). Educational programs help address these challenges by supporting informed decisions (Mariano, Villano, and Fleming 2012). Extension services work best when they promote knowledge exchange, fostering interaction, learning, and innovation (Blackstock et al. 2010).

Increasing supply

Supply-side augmentation involves both unconventional waters, such as water reuse and desalination, and conventional water, such as rainwater harvesting and water storage expansion. In regions already facing severe water scarcity and variability, supply-side augmentation offers an important approach to addressing water challenges.

Water recycling and reuse

Given the finite amount of freshwater, an effective solution is water recycling, which refers to treating already used water for appropriate reuse. By converting municipal or industrial effluents into a new supply, water recycling directly offsets losses from declining precipitation and river flows. Reuse offers three significant benefits: it recovers valuable water, recaptures energy and other scarce resources, and enhances environmental sustainability (Khemka and Eberhard 2025).

Water reuse can take various forms, depending on the source and intended application. For example, municipal wastewater can be recycled for agricultural purposes, such as irrigating crops in Coquimbo, Chile, or for urban landscaping, as practiced in many cities in the western United States. Additionally, water can be reused for drinking purposes. In indirect potable reuse, treated water is first released into the environment before being integrated into the drinking water system, as seen in Singapore's NEWater program. Alternatively, water can be treated in facilities to meet drinking standards, as demonstrated in Windhoek, Namibia. The concept of reusing water to fill water demand aligns with the circular economy approach, which seeks to separate economic growth from negative environmental impacts.

Although water reuse is a powerful strategy for maximizing local water resources, its adoption is often hindered by several barriers. One key challenge is inadequate pricing, which fails to generate enough revenue to cover the full costs of building and maintaining water reuse infrastructure, ultimately discouraging future investment. In regions where water scarcity is not yet severe or where policy frameworks are weak, there is little incentive to prioritize or implement reuse practices. Additionally, negative public perceptions—particularly around direct potable reuse—pose significant obstacles, limiting broader acceptance despite the safety and reliability of modern treatment technologies.

To overcome these challenges, a range of targeted strategies is needed. Policy levers, such as tariffs for used water treatment, abstraction and discharge fees, and water resource taxes, can be designed to better reflect water's true costs and mobilize private capital for infrastructure investments for water reuse. Enhancing quality regulations, adopting technological innovations to achieve cost reduction, and providing consumer education are also important ingredients to make these infrastructures more effective in achieving outcomes.

For instance, in the Islamic Republic of Iran, farmers' acceptance of treated wastewater for irrigation depends on water quality and price; most are willing to use high-quality recycled water, although their willingness to pay is influenced by socioeconomic factors and perceived health risks (Deh-Haghi et al. 2020). An effective example of private capital mobilization is a mining company in Arequipa, Peru, that developed a facility to treat municipal wastewater, retaining a portion of the treated water for its own operations and releasing the rest to a nearby river for other downstream uses, such as irrigation (World Bank 2019b). Through this public-private partnership approach, the city of Arequipa benefited from wastewater treatment at no cost to the taxpayer, and downstream communities benefited from higher-quality water, leading to improved health and agricultural outcomes. When done properly, recycling programs can thus provide a drought-proof source of high-quality water.

Desalination

Desalination is another technology that is gaining traction in expanding the possibility of freshwater availability. Global installed desalination capacity reached roughly 95 million to 97 million m^3 per day by 2020, after growing at an average rate of about 7 percent per year since 2010 (Eke et al. 2020; World Bank 2019a)—more than 5 times the water flow of the Amazon River.

Nearly half of the world's desalinated capacity is located in the Middle East and North Africa (Jones et al. 2019). However, technological advances and declining costs have led to broader appeal, and other regions are quickly catching up. For instance, China has identified seawater desalination as a strategic industry to alleviate its regional water scarcity. It has rapidly expanded its desalination capacity in coastal cities and is increasingly powering new facilities with renewable energy, aligning with its carbon reduction goals (Lin et al. 2021). In Latin America, Chile has increasingly turned to desalination to meet growing water demand. A desalination plant in Antofagasta, Chile, has been established with the goal of providing 100 percent of the city's drinking water (Fitzsimons and Warren 2024).[2]

The technological landscape of desalination is diverse. Traditional methods, such as reverse osmosis, electrodialysis, and multiple-effect distillation, have been widely implemented (Lin et al. 2021). In parallel, efforts to integrate renewable energy sources—such as solar and even geothermal—are gaining traction, promising to reduce both the environmental impact and the operating costs of these plants (Pugsley et al. 2016; Tomaszewska et al. 2021).

However, the viability of desalination for agriculture remains highly context dependent, especially when considering economic trade-offs. Desalinated water is still considerably more expensive than traditional water sources. As such, its use is generally feasible only for high-value crops—such as certain fruits, vegetables, or specialty products—that can absorb the higher cost of irrigation without becoming economically unviable. These conditions are typically met in very water-scarce regions where agricultural markets already support premium pricing, where low-cost renewable energy is abundant, and where the alternative—crop failure or land abandonment—is even more costly.

For the broader agricultural sector, particularly in regions growing lower-value staple crops or operating on thin profit margins, desalination is rarely the most cost-effective solution. In these contexts, demand-side management strategies discussed in the previous section are often more affordable and efficient strategies to reduce water stress.

Although its use in agriculture remains limited today, technological advances and targeted investments are paving the way for broader adoption. In many cases, desalinating brackish water—which contains less salt than seawater—is more cost-effective. Production of high-quality water uses a range of membrane-based technologies, including reverse osmosis, membrane distillation, forward osmosis, and electrodialysis. Innovative hybrid systems, such as combining forward osmosis with electrodialysis, show potential to reduce energy consumption while maintaining efficiency (Suwaileh, Johnson, and Hilal 2020). Furthermore, desalination generates concentrated brine waste that can harm marine ecosystems if not properly managed (Antia 2022). Inland desalination faces further hurdles because of brine disposal impacts on land and groundwater.

Although desalination offers a promising avenue to deal with global water scarcity, overcoming its current economic and environmental hurdles will require continued technological advancement, renewable energy integration, and robust policy support. Looking ahead, establishing clear quality standards, technology transfers, and regulatory frameworks will be the challenges to overcome for effective policy implementation—especially for agricultural use, in which water quality directly influences soil health and crop productivity.

Water storage

Water storage allows for the capture and conservation of excess water during wet periods, ensuring a reliable supply during dry seasons. There is an urgent need for proactive and strategic storage planning that goes beyond built infrastructure alone. Such planning would involve identifying and restoring natural storage systems, such as wetlands and floodplains, that can replenish groundwater, regulate streamflow, and provide reliable local water supplies. In many cases, restoring these systems also benefits downstream users, offering a win-win for water security across entire basins.

More important, effective water resilience requires systemwide planning that integrates both natural and built storage solutions. These solutions include protecting critical natural systems, restoring degraded ecosystems, and building infrastructure where it complements—rather than replaces—nature. Natural ecosystems, particularly intact forests, offer a broader range of provisioning services than managed or artificial systems. They are more effective at sustaining baseflows, filtering water, and buffering extreme events such as floods and droughts.

Water storage interventions aim to sustainably manage water resources within a watershed to enhance water availability and agricultural productivity. Key types include

- Rainwater harvesting, which uses techniques such as planting pits, small ponds, and detention structures to capture and store rainwater;
- Groundwater recharge through check dams and percolation tanks that capture surface runoff to replenish groundwater aquifers, particularly in areas with high runoff potential;
- Spring shed management, which focuses on enhancing the recharge of springs, which are vital water sources in hilly regions;

- Afforestation and reforestation, which involves planting trees to reduce runoff, enhance groundwater recharge, and prevent soil erosion; and
- Catchment management, which combines technologies such as shelterbelts, dams, sand dune stabilization, and gully stabilization to effectively manage water resources within a catchment area.

Conservation agriculture practices such as laser land leveling, minimal tillage, crop rotation, and mulching help conserve soil moisture, prevent erosion, enhance water conservation, and improve soil health. These agricultural methods use stored water to reduce soil disturbance, boost water infiltration, and promote sustainable farming, leading to better crop yields and reduced irrigation dependency. Examples from southern Africa, North America, South America, Central Asia, and South Asia show that in areas with water storage, crop yields increase by 5–235 percent, infiltration rates increase by 64–96 percent, and water consumption is reduced by 40–50 percent (FAO 2016; Mello and van Raij 2006; Siziba et al. 2019; Thierfelder, Bunderson, and Mupangwa 2015).

Groundwater replenishment initiatives and floodplain restoration are key strategies that return water to natural systems and can even improve overall groundwater quality. For example, in arid Sardinia, Italy, a forested infiltration area was developed to improve groundwater quality and recharge by targeting nitrate-vulnerable zones. This policy reduces agricultural nitrate pollution while enhancing aquifer infiltration through forested areas (Carletti et al. 2022). In Germany and the Netherlands, floodplain restoration projects have been scaled up using European Union (EU) sustainability funds to improve water retention, reduce flood risks, and enhance biodiversity. In Qatar, a geographic information system–based framework was implemented to identify areas suitable for artificial aquifer recharge and to recharge these aquifers using treated municipal wastewater. The program necessitated coordination among land management, environmental, and water agencies to address both groundwater depletion and regional water scarcity challenges (Mohieldeen, Elobaid, and Abdalla 2021).

Finally, rehabilitating and maintaining existing infrastructure through sustainable sediment management practices also plays a critical role in expanding storage capacity. Burke et al. (2023) highlight that, in many regions, sedimentation has reduced reservoir capacity by as much as 50 percent, underscoring the urgency of sustainable solutions such as upstream watershed management, dredging, and sediment bypass systems. Effective sediment management not only prolongs the lifespan of reservoirs but also enhances their efficiency, ensuring that storage capacity keeps pace with rising needs.

Improving water allocation

Despite supply- and demand-side interventions, periods of water scarcity will still occur. Effective water allocation policies are necessary to distribute water efficiently and equitably among all users, including marginalized communities and ecosystems, particularly considering the increasing competition from agriculture, urban areas, and industries. Shifting supply and demand also require capacity for reallocation as conditions change.

Water tenure reform

Effective water allocation frameworks incorporate cooperative mechanisms, including the rights to use, transfer, and manage water resources. Although no universal approach exists, a fundamental element across countries is the establishment of strong water sector legislation, which serves as the foundation for defining and protecting water tenure, defined as a set of social relationships between people and groups over water resources, backed by laws, informal institutions, or a blend (Hodgson 2016).

In addition, for water tenure reform to be effective—particularly in supporting demand management—it must be grounded in robust data and monitoring systems, comprehensive planning frameworks, enforceability mechanisms, and a clear definition of rights and responsibilities. Without reliable monitoring and enforcement, tenure systems risk becoming symbolic rather than operational. Strong planning frameworks help integrate water tenure into broader development, climate, and land use strategies, ensuring coherence across sectors. Moreover, enforceability is critical: tenure systems must have legal standing and institutional support to resolve disputes, sanction violations, and uphold entitlements. Often these systems must be designed to be socially equitable.

Successful water tenure reforms involve the principle of fair access to water for all users, including ecosystems and marginalized communities, through legal code or customary systems. Chile was an early adopter of water markets by developing a system in which water rights are treated as private property and are tradeable, meaning that water allocation was determined by the market and water was channeled to high-value uses (Donoso 2018). However, this system has been criticized for enabling indefinite private ownership, contributing to concerns about unequal distribution, and insufficient state control (Macpherson et al. 2023). To address these issues, Chile undertook water code reform in 2022, redefining water as a public good and strengthening government oversight. Key changes included prioritizing human and ecosystem needs over commercial use, introducing time-limited water rights to prevent hoarding, imposing fees on unused rights to discourage speculation, and ensuring fairer access for rural and Indigenous communities. By balancing market mechanisms with stronger regulations, the reforms aim to promote both equitable access and sustainable water management.

Another cost-effective option for water reallocation is through the voluntary sale of water entitlements from users to institutions that manage water in the public interest, for environmental or cultural purposes (Grafton and Wheeler 2018). For nearly two decades, Australia has been undergoing a reform process to use water allocation systems to buy back water resources, mostly expanding these programs to secure entitlements for Aboriginal and Indigenous peoples. Buyback involved the direct purchase of permanent water entitlements from willing sellers. Most of the existing environmental water has been recovered through voluntary offers of water by multiple sellers via an open tender process.

Moreover, the implementation of hybrid water governance models seeks to encompass both formal and customary systems. These models recognize the significance of customary water tenure, wherein Indigenous peoples use traditional practices to manage water resources effectively. The successful integration of these systems relies heavily on a commitment to addressing historical water misallocations and recognizing the important knowledge of water resources held by Indigenous people. This blend of traditional knowledge with formal water governance can lead to more equitable and effective water management systems, ensuring that Indigenous rights are not only upheld but actively incorporated into sustainable development (Hidalgo, Boelens, and Vos 2017; Jackson 2018).

Environmental flows

Environmental water flows, also known as environmental flows, refer to the quantity, timing, and quality of water required to sustain freshwater and estuarine ecosystems, as well as the human well-being that depends on these ecosystems (Arthington et al. 2024). Adequate water flows are necessary to sustain critical habitats, such as spawning grounds for fish, nesting sites for birds, and feeding areas for wildlife. Incorporating environmental water flows into integrated water management policies is essential for achieving sustainable water resource management.

In Chile, the environment is the largest beneficiary of water allocations under the current water tenure scheme. This scheme was strengthened with an ambitious change in the earlier-mentioned water code reform of 2022 that explicitly recognized environmental flows to be returned to natural water bodies.[3] This reform in water rights ensured that water levels in rivers and wetlands remain high enough to preserve biodiversity and increase the buffers against drought periods. Moreover, the law established an update in tariff schemes that recognized the value of healthy ecosystems as providers of essential hydrological services—such as water filtration, habitat support, and flood regulation—that become critical during droughts (Macpherson et al. 2023).

However, in many low- and middle-income countries, environmental water recovery through water tenure systems is often managed under separate governance arrangements that treat environmental and consumptive water uses as distinct and mutually exclusive. This institutional divide has limited the reallocation of water to environmental purposes over time (Horne et al. 2018). Moreover, the effectiveness of environmental water frameworks depends on environmental flow assessments, which require reliable data and robust monitoring systems—elements that are frequently lacking. To ensure long-term environmental benefits, water tenure reforms should be designed with a strong emphasis on developing frameworks supported by coordinated governance and comprehensive data systems.

Transboundary water and interbasin transfers

Transboundary water management and cooperation are increasingly recognized as critical to ensure equitable and reasonable use of shared water resources. Transboundary arrangements and cooperation can foster fair allocation and sharing in both intra- and interbasin contexts, enhancing sustainable water management and preventing conflict (Schmeier and Blumstein 2025).

A fundamental policy of effective transboundary water management is the establishment of robust frameworks that facilitate cooperation among riparian states, including the establishment of joint management mechanisms (for example, treaties) and elaboration of (sub-)basin-level agreements. Reliance on the general international water law principles, such as the duty to cooperate, prevention of significant harm, equitable and reasonable utilization and ecosystem protection, is essential to mitigate disputes and ensure equitable resource distribution (Rahaman 2009, 2012).

Basin-level agreements, furthermore, often include provisions for ecosystem protection and conservation in recognition that riverine ecosystems are fundamental for river health, water quality and availability, and flood and drought mitigation. More recently, the integration of surface and groundwater governance in international agreements has been gaining traction, reinforcing integrated management of shared water resources (Lautze et al. 2018).

Another key policy principle of transboundary water agreements is the inclusion of provisions for data sharing, joint planning, and monitoring to mitigate the risks of water scarcity. These mandates are most often taken on by joint mechanisms established through basin-level agreements. The joint mechanisms provide a platform for sustained cooperation and joint decision-making.

For instance, the Senegal River Basin Organization (Organisation de Mise en Valeur du Fleuve Senegal) was originally established in 1972 as a joint mechanism to address the impacts of severe droughts in West Africa. It has since developed into an organization that owns and manages joint infrastructure assets on behalf of its member countries to increase agriculture productivity, navigation, and energy benefits from the Senegal River, among others. In 2021, its role was further expanded to groundwater when it was mandated to co-host, with the Organization for the Development of the River Gambia, the regional working group elaborating a cooperative management framework for the Senegal-Mauritania Aquifer Basin.

In addition to transboundary cooperation on international rivers, basin transfer schemes within national boundaries are also feasible mechanisms to transfer water from wet to dry areas. The Orange River Project in South Africa facilitates interbasin transfers to support rural livelihoods through agricultural irrigation and hydroelectric power generation. Although the project generated positive socioeconomic impacts, it also led to significant environmental challenges. The unintended consequences, such as invasive species and algal blooms associated with interbasin transfers, are relevant examples that underscore the need to consider ecological impacts in basin planning (Vazquez and Muneepeerakul 2021). Interbasin transfer can also help expand water storage.

Finally, integrating the virtual water trade into water allocation strategies helps mitigate water shortages in water-scarce regions. At the global level, more strategically harnessing the virtual water trade—linking water-rich regions with water-scarce ones—could relieve pressure on depleted resources and support fast-growing economies (Flach et al. 2016; Vallino, Ridolfi, and Laio 2021).

Using cross-cutting policy levers

Effective water governance hinges on a set of cross-cutting levers that are crucial for the implementation of demand, supply, and allocation solutions. These policy levers are described in the following sections. Table 4.1 outlines a phased policy action road map for implementing these proposed policy levers.

TABLE 4.1 Policy roadmap for the water sector to address the continental drying crisis

Policy lever	Policy action: Short term (0–1 year)	Medium term (1–3 years)	Long term (>3 years)
Strengthen institutions	• Identify institutional gaps in water allocation and monitoring. • Launch capacity-building programs for local water agencies. • Develop national water efficiency standards. • Undertake a rapid assessment of environmental water requirements for critical ecosystems.	• Establish interagency coordination mechanisms for IWRM. • Establish regulatory frameworks for groundwater management, including licensing and abstraction limits. • Strengthen decentralized water governance. • Establish clear PPP frameworks to promote private investment in water.	• Promote transboundary water cooperation mechanisms. • Institutionalize adaptive water planning processes informed by data, climate scenarios, and stakeholder engagement.
Reform water tariffs and repurpose subsidies	• Conduct water tariff and subsidy reviews, including assessing implicit subsidies (such as unmetered water, subsidized energy for groundwater) to inform the design of cost-recovery tariffs. • Evaluate the distributional impacts of tariff reforms and assess whether existing social safety nets are sufficient to ensure affordability.	• Launch a public communication campaign to build support for reform. • Pilot (a) volumetric tariffs for agricultural water supply, (b) performance-based tariffs, and (c) repurposing of agricultural water subsidies to support water-saving practices (such as drip irrigation) and payment for environmental services for forest and wetland protection in selected high-use areas. • Enhance social safety net as necessary to ensure affordability.	• Scale up cost-recovery tariff. • Expand performance-based tariff and subsidies linked with conservation outcomes. • Establish a national authority to continuously monitor water pricing and implement cost-recovery tariffs and targeted subsidies.

(continued)

TABLE 4.1 Policy roadmap for the water sector to address the continental drying crisis (*continued*)

Policy lever	Policy action		
	Short term (0–1 year)	**Medium term (1–3 years)**	**Long term (>3 years)**
Adopt water accounting	• Conduct an audit of water use, starting with synthesizing existing data on water consumption and losses. • Determine the amounts of water formally allocated for consumptive and nonconsumptive uses, to identify potential reserves or resources available for environmental flows.	• Launch public data portals that visualize water flows and sectoral usage to build public awareness and accountability. • Develop water-efficiency benchmarks across basins and sectors on the basis of water accounting.	• Institutionalize water accounting to guide water resources planning and allocation.
Leverage data and technology innovation	• Deploy smart meters to enable volumetric pricing and detect overuse. • Pilot the use of downscaled GRACE data for near real-time monitoring of water storage change. • Introduce AI-driven systems to enable detection of leaks and optimization of irrigation scheduling. • Introduce Internet of Things technologies (for example, soil moisture sensors and mobile platforms) to support smallholder farmers' decision-making.	• Institutionalize satellite-based tracking (for example, using downscaled GRACE) of water storage change. • Expand decentralized water decision-making with AI-powered digital controls and sensors.	• Promote in situ water monitoring and data sharing to further enhance the resolution of GRACE data. • Scale precision agriculture using AI-driven irrigation advisory systems, drone spraying, and cloud-based water scheduling. • Integrate real-time allocation models with legal water rights systems, ensuring enforcement and adaptive reallocation during droughts.
Value water in trade	• Quantify the water footprint of key agricultural and industrial exports, especially in water-stressed basins. • Promote public awareness of the water footprint of imports by, for example, creating water intensity labels for imports.	• Establish interministerial platforms (for example, water-agriculture-trade) to coordinate planning and ensure that water is a factor in strategic trade decisions. • Explore technical quotas on virtual water imports based on the sustainability of trade partners' water management.	• Integrate virtual water balances into national food and trade security strategies, identifying optimal import-export mixes to promote water and food security. • Institutionalize water valuation tools in trade negotiations and agreements, particularly in regions reliant on export agriculture (for example, floriculture, rice, or cotton).

Source: Original elaboration for this publication.
Note: AI = artificial intelligence; GRACE = Gravity Recovery and Climate Experiment; IWRM = integrated water resource management; PPP = public-private partnership.

Strengthen institutions

Addressing the continental drying water crisis requires strong institutions to ensure robust regulation, equitable access, and conflict resolution. Strengthening institutions requires several key components:

- Integrated management of water, land, and related resources across various sectors and boundaries, involving multilevel governance at global, regional, national, and subnational levels
- Comprehensive planning
- Enhancement of institutional capacities
- Development of legal and policy frameworks
- Engagement with stakeholders and promotion of social inclusion
- Establishment of conflict resolution and cooperation mechanisms.

Institutions at all levels must work together to improve water resilience. At the local level, community institutions, municipal authorities, and user groups play important roles in managing water demand and adapting to water scarcity. For example, in many countries, the formation of water user associations and local irrigation committees has enhanced the maintenance of irrigation systems and rationing of water during dry spells, resulting in more equitable and efficient water use at the farm level. Enhancing local governance systems through training, improved financing models, and connections to higher-level agencies is therefore important to improve results.

Water institutions provide oversight of water management and governance on a regional scale, encompassing river basins or aquifers shared by multiple administrative areas or nations. Cooperative regional institutions can facilitate water distribution, conflict resolution, and coordinated drought responses across extensive areas. The EU Water Framework Directive has been instrumental in shaping water policy across member states and improving water quality. Nonetheless, more than half of the world's 310 international rivers and nearly all transboundary aquifers lack cooperative agreements. This absence of institutional frameworks represents a significant regional vulnerability during periods of drying (McCracken and Wolf 2019).

At the national level, governments establish the legal and policy framework for water management and coordinate water management across regions. To ensure effective integrated water resources management, a lead national agency with a cross-sectoral mandate can support coordination across ministries and sectors.

Strengthening institutions at all levels—local, national, and regional—is therefore vital for improving water resilience and attracting investment. However, institutional reform often faces resistance from entrenched interests and structural barriers, such as lack of political priority or limited resources to build capacities. Overcoming these challenges requires sustained learning, transparent monitoring, and economic tools that empower communities and build institutional capacities through knowledge sharing and policy coherence (Sojamo et al. 2023). For example, strengthening regulatory agencies and improving data transparency can help mitigate corruption and ensure compliance with water allocation rules (Gupta, Pahl-Wostl, and Zondervan 2013).

Reform water tariffs, and repurpose subsidies

Water pricing and subsidies are important economic instruments that shape water demand, supply, and water tenure systems. When properly designed, they strengthen water sustainability by promoting water conservation and equitable allocation. In contrast, as discussed in chapter 1, water underpricing can lead to overuse and resource depletion.

Although political resistance to higher water prices is common, innovative tariff designs, supported by effective communication campaigns and well-structured compensation mechanisms, can create a more politically viable path forward. Proper compensation mechanisms are crucial to protect vulnerable populations—such as smallholder farmers and low-income households—from negative financial burdens. Studies suggest that water pricing faces less political challenge when the revenue generated is reinvested in social safety nets or alternative livelihood programs to mitigate adverse impacts (Jia et al. 2023).

Successful tariff reforms for the rural water sector in Ghana underscore the importance of coupling pricing adjustments with targeted support measures, clear communication, and institutional accountability to build public trust and ensure equitable outcomes. In Ghana, local water managers in small towns, who are responsible for the rural/community water sector and those falling under the Community Water and Sanitation Agency and municipalities, set tariffs on the basis of affordability and financial sustainability, leading to improved water management practices and cost recovery of operations and maintenance. Higher-income groups faced increased tariffs, with the additional revenue directed toward cost recovery and infrastructure investment. Meanwhile, lower-income populations benefited from subsidized rates, which resulted in unprecedented rapid expansion in access while stabilizing utility finances (Fielmua and Dongzagla 2020).[4]

Moreover, repurposing subsidies can yield a double dividend. For instance, redirecting water and energy subsidies to targeted support for drip irrigation and incentivizing farmers to protect wetlands, forests, and groundwater recharge through payments for environmental services can simultaneously reduce water waste and promote sustainable resource management. Performance-based tariffs—which link water charges to consumption levels and service performance—can create incentives for conservation.

Morocco exemplifies the effectiveness of performance-based tariff systems. In the early 2000s, regional agricultural development agencies introduced performance-based tariffs in large-scale irrigation schemes. These tariffs linked water charges to consumption volumes and infrastructure performance, incentivizing conservation. The reform, along with the adoption of drip irrigation, led to a 30 percent reduction in agricultural water use (Silva-Novoa Sánchez et al. 2022).

By using well-calibrated pricing mechanisms, coupled with thoughtful subsidies, policy makers can foster sustainable water use, mobilize private participation, stimulate technological innovations, and promote resilience among communities. However, to make pricing and subsidies work, strong governance and institutions, as well as water accounting and data tools, are essential for their design, targeting, and effective implementation.

Adopt water accounting

Water accounting is a comprehensive, systematic approach to managing water resources that involves organizing and analyzing data on water use, availability, and quality. One of the key benefits of water accounting is its role in improving decision-making. By offering detailed insights into water use and productivity, this approach enables stakeholders to identify inefficiencies and allocate water more effectively across various sectors (Vardon et al. 2025). This detailed understanding is essential for integrated water resources management, where balancing supply and demand is critical. In addition, addressing water scarcity is another major advantage of water accounting. By identifying areas where water is being depleted and assessing overall availability, this method helps decision-makers anticipate shortages and manage competing demands.

Finally, water accounting is essential for designing tariffs that reflect the true cost of water services and for targeting subsidies effectively. Tariffs designed on the basis of accurate water accounting can promote economic efficiency by ensuring that water utilities recover the costs of service delivery. Water accounting helps to identify the water consumption patterns of different user groups, enabling the design of subsidies that are targeted at the poorest users. This design ensures that subsidies are used effectively without compromising the financial sustainability of water institutions.

Spain uses water accounting to manage its river basin management plans under the EU Water Framework Directive. The water accounting data are recurrently collected to inform dynamic pricing and targeted subsidies to use water in agriculture more efficiently. Water accounting is used to adjust water tariffs on the basis of the severity of scarcity, with higher prices for water-intensive crops. Similarly, water accounting data help define medium-term local investment strategies in communities that use subsidies for modernizing irrigation systems and promoting drought-resistant crops. India's National Water Policy incorporates water accounting to boost efficiency and sustainability (Bassi, Schmidt, and De Stefano 2020). In Rajasthan, water accounting data have been used to design subsidies that support drought-affected farmers and promote water-saving technologies.

Leverage data and technological innovation

Advanced data and technologies offer real-time insights, enabling efficient use and proactive responses to scarcity. This report highlights Gravity Recovery and Climate Experiment (GRACE) data as a game changer in monitoring changes in terrestrial water storage. As demonstrated, integrating downscaled GRACE data with socioeconomic data can help decision-makers identify unsustainable land and water use driving water depletion and pinpoint hot spot regions where water demand is rising, supply is falling, and efficiency is low. Paired with in situ weather and water monitoring, GRACE data can be further downscaled to hyperlocal levels (approximately 1 km) to enable near real-time water monitoring and support precision agriculture.

The convergence of big data, artificial intelligence (AI), and smart technologies can further transform water management by making it more predictive, responsive, and efficient. AI-driven systems now enable real-time monitoring of water quality, detect leaks across infrastructure, and optimize irrigation scheduling on the basis of crop and weather conditions. Internet of Things technologies, such as soil moisture sensors and mobile platforms, empower smallholder farmers in countries such as Kenya and Spain to make data-informed decisions, conserving water and increasing yields. These systems also serve broader water governance goals by improving transparency and equipping institutions with tools for better allocation and accountability. Remote sensing, evapotranspiration tracking, and smart irrigation tools are pushing the boundaries of efficiency. In the United States, localized evapotranspiration data are empowering farmers and resource managers to fine-tune water use with unprecedented accuracy (DeMarco 2024).

Farmer-led water metering and smart irrigation systems are a cornerstone of this digital shift. They not only track water use at the field level but also help resolve disputes over water rights, enhance policy compliance, and enable more effective targeting of subsidies. Techniques such as automated drip and sprinkler systems, when paired with AI-enabled irrigation schedules, have been shown to reduce water use by 30–40 percent while boosting crop productivity. In Türkiye, prepaid water metering has shifted cropping patterns to less water-intensive agriculture while achieving full cost recovery for irrigation systems. These innovations demonstrate how integrating data analytics, AI, and local action can strengthen water allocation systems and institutional capacity, particularly in groundwater governance, where aquifer zoning, licensing, recharge strategies, and real-time monitoring must all be coordinated.

Value water in trade

Trade and finance also offer powerful levers to enhance water efficiency. The inclusion of nontariff measures (NTMs) in trade agreements is beginning to link market access to sustainable water practices. For instance, under the EU's Common Agricultural Policy or the Indo-Japan Partnership Agreement, exporters must meet environmental standards, including those on water use, to qualify for market access. Similarly, the African Continental Free Trade Area promotes NTMs that encourage water-efficient agricultural practices. Although water-related NTMs currently make up only 0.07 percent of all global NTMs, they offer untapped potential. Demand-side NTMs, such as labeling schemes for water-efficient products, empower consumers to support sustainability by choosing products with smaller water footprints. Countries such as Chile are already using these tools to enhance sustainability in export sectors.

Supply-side NTMs recognize the value of water in trade by explicitly defining thresholds for water intensity in traded goods and commodities. Supply-side NTMs include technical quotas for trade, which activate trade restrictions when products fail to meet sustainability standards. Other measures include technical barriers to trade, which relate to product and processing specifications, and food safety requirements. Some countries adopt maximum residue limits, which restrict pesticide and residue levels in crops to protect land and soil quality.

Demand-side NTMs provide consumers with more information about the products they purchase. These NTMs include administrative procedures aimed at progressively reducing demand for certain products or enforcing temporary consumption waivers during specific conditions (for example, drought-sensitive products). Other NTMs inform consumers about sustainable irrigation and farming practices, aligning consumption with long-term water conservation and climate goals. Additionally, certification and labeling schemes for water-efficient products empower consumers to make informed choices based on water footprints and sustainability. Labels that communicate the water story behind a product can inform consumers and markets about the intrinsic value of water sustainability. Just as nutrition labels are essential for human health, water labels can serve a similar purpose for planetary health.

Addressing financing needs

Sustainable water management cannot advance without substantial investment. Bridging the estimated $6.7 trillion financing gap by 2030 (Whiting 2024) will require strong mobilization of private capital. Private investment thrives when rules are clear, revenue models are solid, and public authorities are reliable partners. Tying financial returns to performance outcomes ensures accountability and stronger incentives for private sector participation. However, if water is not priced properly and subsidies are ineffective, investors do not usually see a clear map of expected returns. Data and information are catalyzers of such policies and levers.

In addition, adopting regulatory frameworks that enable more efficient use of financial instruments such as blended finance, green and blue bonds, and public-private partnerships is vital for reducing investment risk and aligning public and private interests. Tax incentives, risk-sharing mechanisms, and guarantees from development finance institutions can further derisk investment and make the sector more attractive. Additionally, water-related insurance products—such as weather-indexed drought insurance schemes—can stabilize returns and increase investor confidence. Together, these tools form a critical part of the solution to the water sector's investment needs.

Conclusion

At the time the great English poet Samuel Taylor Coleridge wrote, "Water, water everywhere, nor any drop to drink," the world was no doubt a different place. Although his observation somehow still rings true today, Coleridge, whose verse not only borrowed from nature but also exploded its boundaries, could probably now open with another line saying, "Land, land everywhere, nor any piece not dry," for in recent years continental drying—the long-term reduction in freshwater availability across large landmasses—has become an alarming trend. This report reveals that, besides greater unpredictability of droughts and deluges, global land is losing freshwater reserves at an accelerated pace.

How can we address the crisis of continental drying? This chapter recommends a three-pronged strategy of managing demand, augmenting water supply, and improving water allocation. Implementing this strategy requires collective action from governments, the private sector, agricultural communities, and water consumers, among others. Implementing advanced technologies, improved management practices, digital transformation, strong governance and institutions, and supportive economic instruments can lead to substantial water savings and efficiency gains to the world. The challenge is immense, and urgent collaboration is essential to securing a sustainable water future.

Notes

1. AWD is more effective when combined with improved varieties, timely planting and fertilization, strong weed control, and good postharvest practices. Because adoption also hinges on effective national irrigation administration and irrigators' association coordination and incentives such as volumetric pricing, AWD is one element of an integrated water and crop management package, not a standalone solution.
2. However, Chile still lacks a comprehensive desalination strategy and a regulatory framework to integrate desalinated water into its overall water systems for human consumption.
3. Practical challenges remain to make this work in practice, such as limited hydrological data, competing water use demands, and the lack of formal basin governance mechanisms.
4. Tariffs for these rural systems usually cover operation, maintenance, and limited repair or expansion costs, but not capital costs.

References

Alauddin, M., M. A. R. Sarker, Z. Islam, and C. Tisdell. 2020. "Adoption of Alternate Wetting and Drying (AWD) Irrigation as a Water-Saving Technology in Bangladesh: Economic and Environmental Considerations." *Land Use Policy* 91: 104430. https://www.sciencedirect.com/science/article/pii/S026483771931868X.

Alcoforado de Moraes, M. M. G., G. N. Souza da Silva, M. P. da Cunha, N. B. M. Dias, T. F. Cardoso, J. J. M. Guilhoto, and R. R. S. Santos. 2021. "Integration of a Network-Based and an Economy-Wide Water Model to Support Decision Making on Water Resources Planning and Management in Northeastern Brazil." *Frontiers in Water* 3: 681723. https://doi.org/10.3389/frwa.2021.681723.

Al-Ghobari, H., and A. Dewidar. 2021. "A Comparative Study of Standard Center Pivot and Growers-Based Modified Center Pivot for Evaluating Uniformity Coefficient and Water Distribution." *Agronomy* 11 (8): 1675. https://doi.org/10.3390/agronomy11081675.

Almeida, A., R. Coelho, J. Costa, and A. Farías. 2017. "Methodology for Dimensioning of a Center Pivot Irrigation System Operating With Dripper Type Emitter." *Engenharia Agrícola* 37 (4): 828–37. https://doi.org/10.1590/1809-4430-eng.agric.v37n4p828-837/2017.

Antia, David J. 2022. "Catalytic Partial Desalination of Saline Water. " *Water* 14 (18): 2893. https://doi.org/10.3390/w14182893.

Arsil, P., Y. Tey, M. Brindal, A. Ardiansyah, and E. Sumarni. 2022. "Perceived Attributes Driving the Adoption of System of Rice Intensification: The Indonesian Farmers' View." *Open Agriculture* 7 (1): 217–25. https://doi.org/10.1515/opag-2022-0080.

Arthington, A. H., D. Tickner, M. E. McClain, M. C. Acreman, E. P. Anderson, S. Babu, C. W. S. Dickens, et al. 2024. "Accelerating Environmental Flow Implementation to Bend the Curve of Global Freshwater Biodiversity Loss." *Environmental Reviews* 32 (3): 387–413. https://doi.org/10.1139/er-2022-0126.

Ashofteh, P., T. Rajaee, P. Golfam, and X. Chu. 2019. "Applying Climate Adaptation Strategies for Improvement of Management Indexes of a River–Reservoir Irrigation System." *Irrigation and Drainage* 68 (3): 420–32. https://doi.org/10.1002/ird.2336.

Aziz, M., S. A. Rizvi, M. A. Iqbal, S. Syed, M. Ashraf, S. Anwer, M. Usman, et al. 2021. "A Sustainable Irrigation System for Small Landholdings of Rainfed Punjab, Pakistan." *Sustainability* 13 (20): 11178. https://doi.org/10.3390/su132011178.

Baidoo, I., K. O. Sam, M. A. Diaba, E. C. Brown-Appiah, Y. P. Akomea, A. E. Atiwoto, D. B. Sarpong, and S. Bolwig. 2024. "Integrating Biochar With Solar Drip Irrigation Technology for Smallholder Vegetable Farming in South-Eastern Ghana Yields Profitable Returns." *Asian Journal of Agricultural Extension, Economics & Sociology* 42 (7): 110–25. https://doi.org/10.9734/ajaees/2024/v42i72522.

Barrett, C. B., A. Islam, M. A. Malek, D. Pakrashi, and U. Ruthbah. 2021. "Experimental Evidence on Adoption and Impact of the System of Rice Intensification." *American Journal of Agricultural Economics* 104 (1): 4–32. https://doi.org/10.1111/ajae.12245.

Bassi, N., G. Schmidt, and L. De Stefano. 2020. "Water Accounting for Water Management at the River Basin Scale in India: Approaches and Gaps." *Water Policy* 22 (5): 768–88. https://doi.org/10.2166/wp.2020.080.

Benedict, S., and H. Hussein. 2019. "An Analysis of Water Awareness Campaign Messaging in the Case of Jordan: Water Conservation for State Security." *Water* 11 (6): 1156. https://doi.org/10.3390/w11061156.

Blackstock, K. L., J. Ingram, R. Burton, K. M. Brown, and B. Slee. 2010. "Understanding and Influencing Behaviour Change by Farmers to Improve Water Quality." *Science of the Total Environment* 408 (23): 5631–8. https://doi.org/10.1016/j.scitotenv.2009.04.029.

Brent, D. A., C. Lott, M. Taylor, J. Cook, K. Rollins, and S. Stoddard. 2020. "What Causes Heterogeneous Responses to Social Comparison Messages for Water Conservation?" *Environmental and Resource Economics* 77 (3): 503–37. https://doi.org/10.1007/s10640-020-00506-0.

Brühl, J., and M. Visser. 2021. "The Cape Town Drought: A Study of the Combined Effectiveness of Measures Implemented to Prevent 'Day Zero.'" *Water Resources and Economics* 34: 100177. https://doi.org/10.1016/j.wre.2021.100177.

Bryan, E., C. Chase, and M. Schulte. 2019. *Nutrition-Sensitive Irrigation and Water Management*. Water Global Practice Guidance Note. Washington, DC: World Bank. https://openknowledge.worldbank.org/entities/publication/ffe6b0d9-b457-5984-8712-3d14da524cc2.

Burke, M., J. M. Tront, K. N. Lyon, W. Rex, M. I. Castera Errea, M. C. Varughese, J. T. Newton, et al. 2023. "What the Future Has in Store: A New Paradigm for Water Storage—Overview for Policy Makers." World Bank, Washington, DC. http://documents.worldbank.org/curated/en/099203002012336127.

Cai, D., M. Shoukat, Y. Zheng, H. Tan, F. Meng, and H. Yan 2022. "Optimizing Center Pivot Irrigation to Regulate Field Microclimate and Wheat Physiology under Dry-Hot Wind Conditions in the North China Plain." *Water* 14 (5): 708. https://doi.org/10.3390/w14050708.

Carletti, A., M. Pirastru, M. Deroma, A. Sessini, G. Ghiglieri, and P. P. Roggero. 2022. "Forested Infiltration Area (FIA) Design in the Arborea Nitrate Vulnerable Zone (Sardinia, Italy)." In *Advances in Science, Technology & Innovation*, edited by V. Naddeo, K. H. Choo, and M. Ksibi, editors, 329–332. Cham, Switzerland: Springer. https://doi.org/10.1007/978-3-031-00808-5_76.

Dahlgreen, J., and A. Parr. 2024. "Exploring the Impact of Alternate Wetting and Drying and the System of Rice Intensification on Greenhouse Gas Emissions: A Review of Rice Cultivation Practices." *Agronomy* 14 (2): 378. https://doi.org/10.3390/agronomy14020378.

Dangar, S., and V. Mishra. 2024. "Groundwater Sustainability in India through Non-Rice-Dominated Cropping Pattern." *PNAS Nexus* 3 (9): pgae378. https://doi.org/10.1093/pnasnexus/pgae378.

Deh-Haghi, Z., A. Bagheri, Z. Fotourehchi, and C. Damalas. 2020. "Farmers' Acceptance and Willingness to Pay for Using Treated Wastewater in Crop Irrigation: A Survey in Western Iran." *Agricultural Water Management* 239: 106262. https://doi.org/10.1016/j.agwat.2020.106262.

DeMarco, E. 2024. "OpenET Study Helps Water Managers and Farmers Put NASA Data to Work." NASA. https://science.nasa.gov/science-research/earth-science/water-energy-cycle/openet-moisture-measurement-tool-is-proving-highly-accurate/.

Donoso, G. 2018. "Overall Assessment of Chile's Water Policy and Its Challenges." In *Water Policy in Chile*, edited by G. Donoso, 209–219. Cham, Switzerland: Springer International. https://doi.org/10.1007/978-3-319-76702-4_14.

Du, J., Y. Laghari, Y.-C. Wei, L. Wu, A.-L. He, G.-Y. Liu, H.-H. Yang, Z.-Y. Guo, and S. J. Leghari. 2024. "Groundwater Depletion and Degradation in the North China Plain: Challenges and Mitigation Options." *Water* 16 (2): 354. https://doi.org/10.3390/w16020354.

Dwipa, I., M. Kasim, N. Rozen, and N. Nurhamidah. 2020. "Land Flooding Effect Before and After Planting on Rice Yield in the System of Rice Intensification." *International Journal on Advanced Science Engineering and Information Technology* 10 (3): 1297–303. https://doi.org/10.18517/ijaseit.10.3.10973.

Eke, J., A. Yusuf, A. Giwa, and A. Sodiq. 2020. "The Global Status of Desalination: An Assessment of Current Desalination Technologies, Plants and Capacity." *Desalination 495:* 114633. https://doi.org/10.1016/j.desal.2020.114633.

Esterhuyse, S., D. Vermeulen, and J. Glazewski. 2019. "Regulations to Protect Groundwater Resources during Unconventional Oil and Gas Extraction Using Fracking." *WIREs Water* 6 (6): e 1382. https://doi.org/10.1002/wat2.1382.

Faisal, M., E. Edwards, and E. Berglund. 2022. "Dynamic Pricing Framework for Water Demand Management Using Advanced Metering Infrastructure Data." *Water Resources Research* 60 (9): e2023WR035246.

FAO (Food and Agriculture Organization of the United Nations). 2000. *Deficit Irrigation Practices*. Water Reports 22. Rome: FAO.

FAO (Food and Agriculture Organization of the United Nations). 2016. *Save and Grow: Maize, Rice, Wheat*. Rome: FAO.

Ferraro, P. J., and M. K. Price. 2013. "Using Nonpecuniary Strategies to Influence Behavior: Evidence from a Large-Scale Field Experiment." *Review of Economics and Statistics* 95 (1): 64–73. https://doi.org/10.1162/REST_a_00344.

Fielmua, N., and A. Dongzagla. 2020. "Independent Water Pricing of Small Town Water Systems in Ghana." *Heliyon* 6 (6): e04299. https://doi.org/10.1016/j.heliyon.2020.e04299.

Fitzsimons, E., and P. Warren. 2024. "Desalination Investment for Copper Mining: Barriers and Opportunities in Chile." *Extractive Industries and Society* 17: 101449. https://doi.org/10.1016/j.exis.2024.101449.

Flach, R., Y. Ran, J. Godar, L. Karlberg, and C. Suavet. 2016. "Towards More Spatially Explicit Assessments of Virtual Water Flows: Linking Local Water Use and Scarcity to Global Demand of Brazilian Farming Commodities." *Environmental Research Letters* 11 (7): 075003. https://doi.org/10.1088/1748-9326/11/7/075003.

Gaspard, N., N. Félicien, M. J. C. Noel, T. J. Claude, M. J. Claude, and D. M. Mburu. 2023. "Impact of Irrigation Systems Use in Agriculture Farming in Rwanda: A Case Study of Kagitumba, Nasho and Muyanza Schemes." *International Journal of Agricultural and Applied Sciences* 4 (1): 1–11. https://doi.org/10.52804/ijaas2023.411.

Grafton, R. Q., J. Williams, C. J. Perry, F. Molle, C. Ringler, P. Steduto, B. Udall, et al. 2018. "The Paradox of Irrigation Efficiency." *Nature Sustainability* 1 (9): 466–72. https://doi.org/10.1038/s41893-018-0138-5.

Grafton, Quentin, and Sarah Ann Wheeler. 2018. "Economics of Water Recovery in the Murray-Darling Basin, Australia." *Annual Review of Resource Economics* 10 (1): 487–510 https://doi.org/10.1146/annurev-resource-100517-023039.

Green Nylen, N., M. Kiparsky, K. Archer, K. Schnier, and H. Doremus. 2017. "Trading Sustainably: Critical Considerations for Local Groundwater Markets under the Sustainable Groundwater Management Act." Wheeler Water Institute, Center for Law, Energy & the Environment, UC Berkeley School of Law. https://www.law.berkeley.edu/research/clee/research/wheeler/trading-sustainably/.

Gupta J., C. Pahl-Wostl, and R. Zondervan. 2013. "'Glocal' Water Governance: A Multi-Level Challenge in the Anthropocene." *Current Opinion in Environmental Sustainability* 5 (6): 573–80. https://doi.org/10.1016/j.cosust.2013.09.003.

Hidalgo, J. P., R. Boelens, and J. Vos. 2017. "De-Colonizing Water: Dispossession, Water Insecurity, and Indigenous Claims for Resources, Authority, and Territory." *Water History* 9 (1): 67–85. https://doi.org/10.1007/s12685-016-0186-6.

Hodgson, S. 2016. "Exploring the Concept of Water Tenure." Food and Agriculture Organization, Rome, Italy. https://openknowledge.fao.org/items/385bf34f-df6c-43f2-b705-07aa41233ab6.

Horne, A. C., E. L. O'Donnell, A. J. Loch, D. C. Adamson, B. Hart, and J. Freebairn. 2018. "Environmental Water Efficiency: Maximizing Benefits and Minimizing Costs of Environmental Water Use and Management." *WIREs Water* 5: e1285. https://doi.org/10.1002/wat2.1285.

Hussainzada, W., and H. S. Lee. 2022. "Effect of an Improved Agricultural Irrigation Scheme with a Hydraulic Structure for Crop Cultivation in Arid Northern Afghanistan Using the Soil and Water Assessment Tool (SWAT)." *Scientific Reports* 12: 5186. https://doi.org/10.1038/s41598-022-09318-2.

Jackson, S. 2018. "Water and Indigenous Rights: Mechanisms and Pathways of Recognition, Representation, and Redistribution." *WIREs Water* 5: e1314. https://doi.org/10.1002/wat2.1314.

Jararweh, Y., S. Fatima, M. Jarrah, and S. AlZu'bi. 2023. "Smart and Sustainable Agriculture: Fundamentals, Enabling Technologies, and Future Directions." *Computers and Electrical Engineering* 110: 108799. https://doi.org/10.1016/j.compeleceng.2023.108799.

Jia, L., R. Xu, Z. Shen, and M. Song. 2023. "Which Type of Innovation Is More Conducive to Inclusive Green Growth: Independent Innovation or Imitation Innovation?" *Journal of Cleaner Production* 406: 137026. https://doi.org/10.1016/j.jclepro.2023.137026.

Jones, E., M. Qadir, M. Van Vliet, V. Smakhtin, and S. Kang. 2019. "The State of Desalination and Brine Production: A Global Outlook." *Science of the Total Environment* 657: 1343–56. https://doi.org/10.1016/j.scitotenv.2018.12.076.

Khemka, R., and R. Eberhard. 2025. *Scaling Water Reuse: A Tipping Point for Municipal and Industrial Use*. Washington, DC: World Bank. http://documents.worldbank.org/curated/en/099052025124041274.

Lautze, J., B. Holmatov, D. Saruchera, and K. G. Villholth. 2018. "Conjunctive Management of Surface and Groundwater in Transboundary Watercourses: A First Assessment." *Water Policy* 20 (1): 1–20. https://doi.org/10.2166/wp.2018.033.

Levidow, L., D. Zaccaria, R. Maia, E. Vivas, M. Todorovic, and A. Scardigno. 2014. "Improving Water-Efficient Irrigation: Prospects and Difficulties of Innovative Practices." *Agricultural Water Management* 146: 84–94. https://doi.org/10.1016/j.agwat.2014.07.012.

Li, F., W. Yan, Y. Zhao, and R. Jiang. 2021. "The Regulation and Management of Water Resources in Groundwater Over-Extraction Area Based on ET." *Theoretical and Applied Climatology* 146 (1-2): 57–69. https://doi.org/10.1007/s00704-021-03713-x.

Lin, S., H. Zhao, L. Zhu, T. He, S. Chen, C. Gao, and L. Zhang. 2021. "Seawater Desalination Technology and Engineering in China: A Review." *Desalination* 498: 114728. https://doi.org/10.1016/j.desal.2020.114728.

Macpherson, E., C. Clark, P. W. Salazar, N. Baird, A. Akhtar-Khavari, and E. Challies. 2023. "Evolving Rights to (and of) Water in Chile: A Case for Relationship-Based Water Law and Governance." *International Journal of Human Rights* October: 1–26. https://www.tandfonline.com/doi/full/10.1080/13642987.2023.2266719.

Marcos-Garcia, P., M. Pulido-Velázquez, C. S. Ibor, M. García-Mollá, M. Ortega-Reig, A. García-Prats, and C. Girard. 2023. "From Local Knowledge to Decision Making in Climate Change Adaptation at Basin Scale: Application to the Jucar River Basin, Spain." *Climatic Change* 176 (4): 38. https://doi.org/10.1007/s10584-023-03501-8.

Mariano, M. J., R. Villano, and E. Fleming. 2012. "Factors Influencing Farmers' Adoption of Modern Rice Technologies and Good Management Practices in the Philippines." *Agricultural Systems* 110: 41–53. https://doi.org/10.1016/j.agsy.2012.03.010.

Marston, L., and X. Cai. 2016. "An Overview of Water Reallocation and the Barriers to Its Implementation." *Wiley Interdisciplinary Reviews: Water* 3 (5): 658–77.

McCracken, M., and A. T. Wolf. 2019. "Updating the Register of International River Basins of the World." *International Journal of Water Resources Development* 35 (5): 732–82. https://doi.org/10.1080/07900627.2019.1572497.

Meinzen-Dick, R., and C. Ringler. 2008. "Water Reallocation: Drivers, Challenges, Threats, and Solutions for the Poor." *Journal of Human Development* 9 (1): 47–64. https://doi.org/10.1080/14649880701811393.

Mello, I., and B. van Raij. 2006. "No-Till for Sustainable Agriculture in Brazil." *World Association of Soil and Water Conservation Proceedings* 1: 49–57.

Mi, Q., X. Li, X. Li, G. Yu, and J. Gao. 2021. "Cotton Farmers' Adaptation to Arid Climates: Waiting Times to Adopt Water-Saving Technology." *Agricultural Water Management* 244: 106596. https://doi.org/10.1016/j.agwat.2020.106596.

Mitchell, J., A. Shrestha, J. Hollingsworth, D. Munk, K. Hembree, and T. Turini. 2016. "Precision Overhead Irrigation Is Suitable for Several Central Valley Crops." *California Agriculture* 70 (2): 62–70. https://doi.org/10.3733/ca.v070n02p62.

Mohieldeen, Y. E., E. A. Elobaid, and R. Abdalla. 2021. "GIS-Based Framework for Artificial Aquifer Recharge to Secure Sustainable Strategic Water Reserves in Qatar Arid Environment Peninsula." *Scientific Reports* 11 (1): 18184. https://doi.org/10.1038/s41598-021-97593-w.

Molle, François. 2003. "*Development Trajectories of River Basins: A Conceptual Framework.*" Research Paper No. 72, International Water Management Institute, Colombo, Sri Lanka.

Nelson, A., R. Waßmann, B. O. Sander, and L. K. Palao. 2015. "Climate-Determined Suitability of the Water Saving Technology 'Alternate Wetting and Drying' in Rice Systems: A Scalable Methodology Demonstrated for a Province in the Philippines." *PLOS One* 10 (12): e0145268. https://doi.org/10.1371/journal.pone.0145268.

Neogi, M. G., A. M. Uddin, M. T. Uddin, and M. A. Hamid. 2018. "Alternate Wetting and Drying (AWD) Technology: A Way to Reduce Irrigation Cost and Ensure Higher Yields of Boro Rice." *Journal of the Bangladesh Agricultural University* 16 (1): 1–4. https://doi.org/10.3329/jbau.v16i1.36471.

O'Donnell, E. L., and D. E. Garrick. 2019. "The Diversity of Water Markets: Prospects and Perils for the SDG Agenda." *WIREs Water* 6 (5): e1368. https://doi.org/10.1002/wat2.1368.

Ogata, R., S. Segawa, S. Rashid, and H. Nakayama. 2021. "Revenue Augmentation through Improved Water Supply Services: A Case Study of the Smart-Wasa Team of Faisalabad, Pakistan." *Journal of Water, Sanitation and Hygiene for Development* 11 (6): 1097–101. https://doi.org/10.2166/washdev.2021.064.

Peth, D., O. Mußhoff, K. Funke, and N. Hirschauer. 2018. "Nudging Farmers to Comply with Water Protection Rules—Experimental Evidence from Germany." *Ecological Economics* 152: 310–21. https://doi.org/10.1016/j.ecolecon.2018.06.007.

Pfeiffer, L., and C.-Y. C. Lin. 2014. "Does Efficient Irrigation Technology Lead to Reduced Groundwater Extraction? Empirical Evidence." *Journal of Environmental Economics and Management* 67 (2): 189–208. https://doi.org/10.1016/j.jeem.2013.12.002.

Pugsley, A., A. Zacharopoulos, J. Mondol, and M. Smyth. 2016. "Global Applicability of Solar Desalination." *Renewable Energy* 88: 200-219. https://doi.org/10.1016/J.RENENE.2015.11.017.

Rahaman, M. M. 2009. "Principles of Transboundary Water Resources Management and Ganges Treaties: An Analysis." *International Journal of Water Resources Development 25* (1): 159–173. https://doi.org/10.1080/07900620802517574.

Rahaman, M. M. 2012. "Principles of Transboundary Water Resources Management and Water-Related Agreements in Central Asia: An Analysis." *International Journal of Water Resources Development 28* (3): 475–491. https://doi.org/10.1080/07900627.2012.684311.

Richter, B. D., Y. Ao, G. Lamsal, D. Wei, M. Amaya, L. Marston, and K. F. Davis. 2023. "Alleviating Water Scarcity by Optimizing Crop Mixes." *Nature Water* 1 (12): 1035–47. https://doi.org/10.1038/s44221-023-00155-9.

Schmeier, S., and S. Blumstein. 2025. "The Role of Basin Water Treaties and Basin Organizations in Managing Transboundary Water Resources: Taking Stock of Current Practice." In *The Role of Formal and Informal Institutions in Managing Transboundary Basins,* edited by N. A. Zawahri, 51–77. Vol. 3 of *World Scientific Handbook of Transboundary Water Management: Science, Economics, Policy and Politics,* editor-in-chief S. Dinar. Singapore: World Scientific

Silva-Novoa Sánchez, L. M., L. Bossenbroek, J. Schilling, and E. Berger. 2022. "Governance and Sustainability Challenges in the Water Policy of Morocco 1995–2020: Insights From the Middle Draa Valley." *Water* 14 (18): 2932. https://doi.org/10.3390/w14182932.

Siziba, S., K. Nyikahadzoi, C. Makate, and N. Mango. 2019. "Impact of Conservation Agriculture on Maize Yield and Food Security: Evidence from Smallholder Farmers in Zimbabwe." *African Journal of Agricultural and Resource Economics* 14 (2): 89–105.

Sojamo, S., M. Keulertz, J. F. Warner, and J. A. Allan. 2023. "Virtual Water Hegemony: The Role of Agribusiness in Global Water Governance." In *The Role of Agribusiness in Global Water Governance,* edited by Christopher A. Scott and Bernard de Gouvello, 82–169. London: Routledge.

Stitzel, B., and C. Rogers. 2023. "No Tips Please: Informational Messaging for Water Conservation." Preprint, submitted November 20, 2023. https://doi.org/10.21203/rs.3.rs-3554703/v1.

Suwaileh, W., D. Johnson, and N. Hilal. 2020. "Membrane Desalination and Water Re-Use for Agriculture: State of the Art and Future Outlook." *Desalination* 491: 114559. https://doi.org/10.1016/j.desal.2020.114559.

Thierfelder, C., W. Bunderson, and W. Mupangwa. 2015. "Evidence and Lessons Learned from Long-Term On-Farm Research on Conservation Agriculture Systems in Communities in Malawi and Zimbabwe." *Environments* 2 (3): 317–37. https://doi.org/10.3390/environments2030317.

Timotewos, M. T., and M. Barjenbruch. 2024. "Examining the Prospects of Residential Water Demand Management Policy Regulations in Ethiopia: Implications for Sustainable Water Resource Management." *Sustainability* 16 (13): 5625. https://doi.org/10.3390/su16135625.

Timotewos, M. T., M. Barjenbruch, and B. M. Behailu. 2023. "Setting a Baseline Residential Water Demand Management Solution in Urban Towns of Ethiopia." *Water* 15 (5): 930. https://doi.org/10.3390/w15050930.

Tomaszewska, B., G. Akkurt, M. Kaczmarczyk, W. Bujakowski, N. Keles, Y. Jarma, A. Baba, M. Bryjak, and N. Kabay. 2021. "Utilization of Renewable Energy Sources in Desalination of Geothermal Water for Agriculture." *Desalination* 513: 115151. https://doi.org/10.1016/J.DESAL.2021.115151.

Tong, L. A., M. Ulubaşoğlu, and C. Guven. 2022. "Growing More Rice With Less Water: The System of Rice Intensification and Water Productivity in Vietnam." *Australian Journal of Agricultural and Resource Economics* 66 (3): 581–611. https://doi.org/10.1111/1467-8489.12477.

Vallino, E., L. Ridolfi, and F. Laio. 2021. "Trade of Economically and Physically Scarce Virtual Water in the Global Food Network." *Scientific Reports* 11: 22806. https://doi.org/10.1038/s41598-021-01514-w.

Vardon, M. J., T. Lien Le, R. Martinez-Lagunes, O. Batlokwa Pule, S. Schenau, S. R. May, and Q. Grafton. 2025. "Accounting for Water: A Global Review and Indicators of Best Practice for Improved Water Governance." *Ecological Economics* 227: 108396. https://doi.org/10.1016/j.ecolecon.2024.108396.

Vazquez, K., and R. Muneepeerakul. 2021. "Modeling Resilience and Sustainability of Water-Subsidized Systems: An Example from Northwest Costa Rica." *Sustainability* 13 (4): 2013. https://doi.org/10.3390/su13042013.

Wardiha, M. W., I. Istiarto, and F. Nurrochmad. 2021. "Acceptance Analysis of PDAM Sleman's Water Tariff, Based on Revenue and Willingness to Pay Projection." *Journal of the Civil Engineering Forum* 7 (2): 155–64. https://doi.org/10.22146/jcef.60536.

Whiting, K. 2024. "Why Investment in Water Is Crucial to Tackling the Climate Crisis." World Economic Forum, June 5, 2024. https://www.weforum.org/stories/2024/06/investment-water-climate-crisis-sdgs/.

Wight, C., B. D. Richter, G. Ayala, L. Blevins, K. Carter, S. Carter, B. Coughlin, et al. 2025. "Taking Stock of Caps on Water Use: Fostering Sustainability or Falling Short?" *Water Policy* 27 (1): 17–39. https://doi.org/10.2166/wp.2024.071.

World Bank. 2019a. "The Role of Desalination in an Increasingly Water-Scarce World." World Bank, Washington, DC. https://hdl.handle.net/10986/31416.

World Bank. 2019b. "Wastewater: From Waste to Resource—The Case of Arequipa, Peru." World Bank, Washington, DC. https://hdl.handle.net/10986/33110.

Worthmann, C., and S. Esterhuyse. 2022. "A Mobile Application to Protect Groundwater during Unconventional Oil and Gas Extraction." *Royal Society Open Science* 9 (9): 220221. https://doi.org/10.1098/rsos.220221.

www.ingramcontent.com/pod-product-compliance
Lightning Source LLC
LaVergne TN
LVHW070624061126
851857LV00002B/11
9781464822698